T0331019

"In the realm of science and engineering symmetry often plays a central role, determining the form and behavior of structures. This book presents rigorous mathematical description of several types and examples of structural symmetry.

Chapter 1 is an introduction to the symmetry in Japanese art - from precise creases of origami to the elaborate patterns of bamboo weaving and the intriguing balance of toys such as the Yajirobee.

Here the reader is given a tangible connection between everyday experience and the abstract world of symmetry.

Chapter 2 introduces rigorous foundations of the group theory, showcasing how the seemingly simple concept of symmetry can be mathematically described, classified, and represented. Here the reader is provided with mathematical tools necessary for full understanding of the subsequent chapters.

Chapter 3 merges abstract concepts with practical examples, presenting group theory in the context of structural mechanics.

Chapter 4 provides an in-depth look at the Finite Element Method, a powerful tool for engineers and scientists.

The following parts of the book illustrate the applications of group theory to real-world symmetric structures, illustrated with several examples from purely academic to industrial applications.

This book is intended for a wide range of readers: for students of science and engineering who will find meaningful connections between fundamental concepts and practical applications; for engineers and chemists, whether in academia or in industry, who will find it as a valuable source of insights for shape research and in-form design. Last but not least, the book serves as an inspiration and guide for anyone who is curious about the myriad of ways in which symmetry is present in our world. Embrace the mathematical rhythms that underpin the symmetrical wonders around us!"

- **Professor Mirosław J. Skibniewski,** *University of Maryland, College Park, United States of America*

"This book, which focuses on 'Symmetry' - the singular intersection of mathematics and physics, promises to open doors to new academic worlds for its readers. Not only does it delve deeply into specialized concepts such as group theory and finite element methods, but it also reveals the mathematical elements hidden in ordinary items like traditional bamboo crafts and children's toys. By picking up this book, you will notice the mathematical charm behind the things around you. From young students to professionals, this book offers fresh knowledge and perspectives to all its readers. It's bound to enrich your interdisciplinary exploration journey. I recommend you to take this book in your hands and take the first step into the unknown world."

- **Professor Masatoshi Nakazawa,** *Tohoku-gakuin University*

Application of Group Theory to Symmetric Structures

Ario and Zawidzki show readers how to handle symmetric structures in engineering using group-theoretic bifurcation theory as a mathematical tool for the finite element analysis of symmetric structures.

They guide the reader from the initial mathematical concepts through to application examples. Readers will gain a solid theoretical grounding in group theory and strong working knowledge of the use of computational frameworks for structural analysis using mathematical representations of symmetry and physical symmetry. First, the authors elaborate an outline of symmetric structures in engineering and then describe the representation of symmetry and group theory. They then discuss block diagonalization theory and finite element analysis models. This provides readers with the base knowledge needed for Chapter 6, which is based on numerical analysis examples of invariant, static FEM model systems and dynamic model systems of the dihedral group. This unique approach is a vital method that will enable readers to reduce the time and computation needed for accurate analysis so that they can better design such structures. The focus on finite element methods and practical examples and case studies throughout provides a strong practical foundation for anyone studying or working in this field.

The book is a valuable resource for undergraduate and postgraduate students on various courses such as civil and mechanical engineering, architecture, structural engineering, applied mathematics, and physics. Additionally, it describes vital practical solutions for structural engineers, structural system manufacturers, fabricators of prefabricated elements, developers of computational mechanics, and so on.

Ichiro Ario is an assistant professor at the Graduate School of Advanced Science and Engineering, Hiroshima University. He obtained his Doctor of Engineering degree from Nagaoka University of Technology in Japan.

Machi Zawidzki is an assistant professor in the Department of Intelligent Technologies, Institute of Fundamental Technological Research of the Polish Academy of Sciences. He obtained his Doctor of Engineering degree from Ritsumeikan University in Japan.

Application of Group Theory to Symmetric Structures

Ichiro Ario and Machi Zawidzki

CRC Press
Taylor & Francis Group
Boca Raton London New York

CRC Press is an imprint of the
Taylor & Francis Group, an **informa** business

Designed cover image: a deployable/foldable bridge is drawn by I. Ario in Structural Laboratory in Hiroshima University, as Mobile bridge 4.0 drawing

First edition published 2024
by CRC Press
2385 NW Executive Center Drive, Suite 320, Boca Raton FL 33431

and by CRC Press
4 Park Square, Milton Park, Abingdon, Oxon, OX14 4RN

CRC Press is an imprint of Taylor & Francis Group, LLC

ISBN: 978-1-032-67017-1 (hbk)
ISBN: 978-1-032-67037-9 (pbk)
ISBN: 978-1-032-67038-6 (ebk)

DOI: 10.1201/9781032670386

Typeset in CMR10
by KnowledgeWorks Global Ltd.

Publisher's note: This book has been prepared from camera-ready copy provided by the authors.

Contents

Foreword

It is with great pleasure that I write this foreword for the book "Application of Group Theory to Symmetric Structures" edited by Andrew Stow and Kasturi Ghosh; Book Commissioning Editor in CRC Press. This book represents a significant contribution to the field of structural analysis, focusing on the application of group theory in understanding and analyzing symmetric structures.

In recent years, there has been an increasing recognition of the importance of symmetry in various fields, including physics, chemistry, biology, and materials science. Symmetry plays a fundamental role in understanding the behavior and properties of natural and engineered systems. The application of group theory, with its mathematical framework for describing and analyzing symmetry, has proven indispensable in these endeavors.

"Application of Group Theory to Symmetric Structures" is a comprehensive compilation of knowledge and techniques in this field. The book covers a wide range of topics, starting with the foundational concepts of group theory and its relevance to symmetry. It then delves into the specific applications of group theory in the analysis of symmetric structures, including the use of block diagonalization theory, parallel Cholesky decomposition, and numerical efficiency evaluations.

One of the strengths of this book lies in its balance between theoretical discussions and practical applications. The authors have provided clear explanations of complex concepts and techniques, making them accessible to readers with varying levels of expertise. Moreover, the inclusion of numerous examples and case studies enhances the practicality and relevance of the material.

We believe that this book will be of great value to students, researchers, and practitioners in the field of structural analysis. Whether one is interested in the fundamental principles of group theory or seeking practical solutions for analyzing symmetric structures, this book offers a wealth of knowledge and insights. We would like to extend my sincere appreciation to the editors and authors who have contributed to this book. Their expertise and dedication have made this publication possible. We also commend the editorial team for their meticulous work in organizing and presenting the material in a coherent and comprehensive manner.

In conclusion, "Application of Group Theory to Symmetric Structures" is an important resource that bridges the gap between theory and practice in the field of structural analysis. We have no doubt that readers will find this book to be a valuable asset in their pursuit of knowledge and advancements in this fascinating area of study.

Ichiro Ario and Machi Zawidzki

Preface

This book serves as a comprehensive guide to the application of block diagonalization theory in the finite element analysis of symmetric structures, with a focus on numerical efficiency evaluation of parallel Cholesky decomposition for stiffness matrices and group products.

In recent years, block diagonalization theory has been proposed as a coordinate transformation method for finite element analysis of symmetric structures based on group-theoretic bifurcation theory, and its advantages in numerical analysis have been shown. This theory diagonalizes the block matrix by performing coordinate transformations based on the geometric symmetry of the system for various matrices such as the (tangent) stiffness matrix, attenuation, and the mass matrix of symmetric structures. The governing equation is decomposed by this theory into multiple independent equations. Since this coordinate transformation can decompose the governing equation of a symmetric structure represented by a general Cartesian coordinate system as multiple independent equations corresponding to irreducible representations, in a large-scale discrete system structure that requires a huge amount of calculation, it is highly advantageous.

This book summarizes block diagonalization theory for static and dynamic problems in computerized structural analysis. The structure of this work focuses on the basic concepts for using group theory from Chapter 1 to Chapter 3, and in Chapter 4, the main points of the finite element method are summarized.

Chapter 1 describes the background to the development of block diagonalization theory of symmetric structures and its usefulness and validity.

Chapter 2 describes the basic concept of groups and the representation theory of groups for mathematically describing symmetry. The description method and calculation method of symmetry for elucidating the mechanical behavior of a system with geometric symmetry will be described. From such a group-theoretic point of view, a useful physical prediction can be obtained even for a system in which the symmetry of the structural system is not perfect and is somewhat incomplete.

Chapter 3 describes block diagonalization theory derived from Schur's lemma. That is, considering the invariant of the governing equations such as the equation of motion and the equilibrium equation derived from the variational principle of energy, when this conditional equation is satisfied, the governing equation can be block-diagonalized.

In Chapter 4, in order to utilize the block diagonalization theory, we will take up a finite element method model with various discretized elements used in the structural analysis of structures and introduce each mechanical model.

In Chapter 5, while block diagonalization theory holds for all groups, in this chapter, the focus is on the dihedral group D_n, which represents the symmetry of the axi-symmetric structural system, and its irreducible representation. The definition is discussed in detail and, in addition, as an application of block diagonalization theory, this method will be described for the D_n invariant structural system balance equation, differential equation, complex stiffness matrix, and control system matrix. In addition, the evaluation of theoretical computational efficiency related to eigenvalue analysis and the interrelationship between displacement and rotation in symmetry transformation will be examined in detail.

In Chapter 6, we took up abundant numerical analysis examples of D_n invariant axi-symmetric structures, compared them with conventional analysis results, and investigated their usefulness and computational effects in detail.

Chapter 7, titled "Numerical Efficiency Evaluation of Parallel Computing Method," presents a detailed exploration of the numerical efficiency evaluation of parallel Cholesky decomposition for stiffness matrices using the block diagonalization method. This approach utilizes block diagonalization to transform stiffness matrices into diagonal form, enabling efficient numerical analysis. The chapter thoroughly investigates the effects and efficiency of parallel Cholesky decomposition based on the block diagonalization method, utilizing numerical analysis results.

Chapter 8, titled "Products of Group Representation," focuses on the concept related between group products and the finite element method. It explores the properties and applications of group products, which play a vital role in group theory by representing the operation of elements within a group. The chapter discusses the application of group products in symmetric structures and highlights their significance in various fields.

Together, these chapters form a comprehensive exploration of the application of block diagonalization theory in the analysis of symmetric structures. By delving into numerical efficiency evaluation and group products, readers gain a deeper understanding of the benefits and possibilities offered by the block diagonalization method.

This book is intended to serve as a valuable resource for students, researchers, and professionals interested in the application of block diagonalization theory in structural analysis. Our hope is that it provides readers with new knowledge and insights, enabling them to effectively apply this theory in their research and practical endeavors.

Editorial Team Andrew Stow and Kasturi Ghosh, Book Commissioning Editor in CRC Press

Symbols

1. Symbols for groups

C_n, C_j	· · · · · · · · ·	Rotation group (Cyclic group)
D_n	· · · · · · · · ·	Dihedral group for order n
E, e	· · · · · · · · ·	Identity conversion
G	· · · · · · · · ·	Group
$\|G\|$	· · · · · · · · ·	Number of groups
G_{sub}	· · · · · · · · ·	Subgroup
H	· · · · · · · · ·	Coordinate transformation matrix
$R(g), \mathcal{R}(g)$	· · · · · · · · ·	Total number for conversion source g
R	· · · · · · · · ·	Transpose matrix by the rotation operation
S	· · · · · · · · ·	Transpose matrix by the mirroring operation
$T(g)$	· · · · · · · · ·	Irreducible representation matrix for g
U	· · · · · · · · ·	Unitary matrix
a	· · · · · · · · ·	Multiplicity
g	· · · · · · · · ·	Group element
p	· · · · · · · · ·	Total amount of computation for eigenvalue analysis
$q(G)$	· · · · · · · · ·	Total number of blocks in group G
n	· · · · · · · · ·	Order (order of symmetry)
r	· · · · · · · · ·	Rotation conversion, rotation operation
s	· · · · · · · · ·	Mirror conversion, mirror operation
$\mu = (1, j), (2, j)$	· · · · · · · · ·	Irreducible representation
$(\cdot)^\mu, (\cdot)^{(d,j)}$	· · · · · · · · ·	Variables for irreducible representation
χ	· · · · · · · · ·	Matrix index

2. Symbols for structural Mechanics

A	$\cdots\cdots\cdots$	Control system matrix
B	$\cdots\cdots\cdots$	Matrix half-bandwidth, or Strain matrix
C	$\cdots\cdots\cdots$	Damping matrix
D	$\cdots\cdots\cdots$	Material composition matrix
E	$\cdots\cdots\cdots$	Elastic modulus
$\boldsymbol{F}$	$\cdots\cdots\cdots$	Governing equilibrium equation
G	$\cdots\cdots\cdots$	Shear modulas
H	$\cdots\cdots\cdots$	Transpose matrix
J	$\cdots\cdots\cdots$	Jacobian matrix, or Tangent stiffness matrix
K	$\cdots\cdots\cdots$	Stiffness matrix
L	$\cdots\cdots\cdots$	Lagrangian
M	$\cdots\cdots\cdots$	Mass matrix
N	$\cdots\cdots\cdots$	Matrix size, or Shape function
O	$\cdots\cdots\cdots$	Zero matrix
$\boldsymbol{P},\,\boldsymbol{p}\,,\boldsymbol{f}$	$\cdots\cdots\cdots$	Load vector
$S(\lambda)$	$\cdots\cdots\cdots$	Complex dynamic stiffness matrix
$\mathcal{U}$	$\cdots\cdots\cdots$	Strain energy
$\mathcal{W}$	$\cdots\cdots\cdots$	External potential energy, Work
Π	$\cdots\cdots\cdots$	The total potential energy
$\boldsymbol{u}$	$\cdots\cdots\cdots$	Displacement vector
$\boldsymbol{v}$	$\cdots\cdots\cdots$	Translational displacement vector
$\boldsymbol{w}$	$\cdots\cdots\cdots$	Displacement vector after block diagonalization
$\boldsymbol{\theta}$	$\cdots\cdots\cdots$	Rotational displacement vector
λ	$\cdots\cdots\cdots$	Eigenvalues, natural frequency, or Lame's constants
ν	$\cdots\cdots\cdots$	Poisson's ratio
$(\cdot)^i$	$\cdots\cdots\cdots$	Variable for node number i
$(\cdot)_v$	$\cdots\cdots\cdots$	Variables for translational displacement
$(\cdot)_X,(\cdot)_Y,(\cdot)_Z$	$\cdots\cdots\cdots$	X,Y,Z-direction displacement
$(\cdot)_{XY}$	$\cdots\cdots\cdots$	Variables for displacement of the XY plane
$(\cdot)_\theta$	$\cdots\cdots\cdots$	Variables for rotational displacement
$(\cdot)_e,(\cdot)^e$	$\cdots\cdots\cdots$	Matrix variable for element e
$(\cdot)^\mu,(\cdot)^{(d,j)}$	$\cdots\cdots\cdots$	Irreducible representation
Γ_j	$\cdots\cdots\cdots$	Alternative matrix of M,C,K etc.
$\Sigma(\cdot)$	$\cdots\cdots\cdots$	A group representing the symmetry of the column vector of matrices
$\widetilde{\cdot}$	$\cdots\cdots\cdots$	Variables after coordinate conversion

3. Symbols for general notations

C	·········	Complex number field
$\boldsymbol{R}, \mathbf{R}$	·········	Rational number field
V	·········	Volume
$f(\cdot)$	·········	Map conversion
I	·········	Identity matrix
$(\cdot)^{\mathrm{T}}$	·········	Transposed matrix
$(\cdot)^{-1}$	·········	Inverse matrix
$\det\vert\ \vert, \vert\ \ \vert$	·········	Determinant
$\mathrm{diag}[\ ]$	·········	Diagonal matrix
$\mathrm{Re}(\cdot)$	·········	The real part of a complex number
$\mathrm{Im}(\cdot)$	·········	Imaginary part of complex number
$\gcd(a, b)$	·········	The greatest common divisor of a and b
$\mathrm{Tr}(\cdot)$	·········	Operator that adds the diagonal terms of a matrix
δ_{ij}	·········	Kronecker's delta
$\oplus$	·········	Direct sum symbols

Acknowledgment

Authors would like to express sincere gratitude to: Professor Khaji Naser for help in editing part of the manuscript, Assistant Professor Nguyen Huu May and the secretary, Mises Ruiko Ogiwara for checking it; the master students: Mr. Keigo Yoshida, Mr. Ma Don and Mr. Ma Kaijo in Hiroshima University for help in creating illustrations and editing the manuscript using LaTeX on the overleaf system.

This book is a part of a project titled: Arm-Z: an extremely modular hyperredundant low-cost manipulator - development of control methods and efficiency analysis funded by OPUS 17 research grant No. 2019/33/B/ST8/02791 supported by the National Science Centre, Poland.

Part I

This is Fundamental Part

1

Introduction

This book delves into the mysteries and mathematical laws of symmetry in nature and human-made forms. Through the interconnected realms of mathematics, structural mechanics, and computational engineering, it reveals how symmetry influences our understanding, design, and everyday life. From group representation theory to the stability of architectural structures, and complex computer simulations, symmetry serves as a bridge between mathematical elegance and practical application. In this introduction, we explore how symmetry is utilized to comprehend the balance between order and disorder in the natural world and to optimize the efficiency and functionality of man-made objects. Readers will gain insights into the role of symmetry in the forefront of modern science and how it shapes our world.

1.1 A Review of Science and Engineering Approaches for Symmetry

First, symmetry is prevalent in nature, from microscopic entities such as elementary particles, atoms, molecules, and crystal structures to large-scale artificial structures like domes and shells. Research exploring the fundamental characteristics of these symmetrical structures is primarily conducted in the fields of quantum mechanics and quantum chemistry. Nowadays, various sectors are attempting to harness the benefits of symmetry. Group theory has become the established method for mathematically expressing symmetry.

We should describe the outline of this book: it summarizes structural symmetry and structural analysis methods, we consider how to apply them to the structural analysis of the finite element method and describe the analysis method of the structural model that is symmetrically discrete elements.

Attempts to utilize the geometrical symmetry of structures for the finite element method, etc., are now widely carried out. For example, the substructure synthesis method [1, 2], the super-element method [3–6], and the method for expanding finite elements into Fourier series by utilizing axi-symmetry [7–9] have been adapted for large-scale symmetric structures in which folded symmetry is utilized and divided into several regions and then analyzed for each substructure system. However, the description and utilization of symmetry are

DOI: 10.1201/9781032670386-1

carried out semi-empirically in the field of finite element method and structural dynamics, and the treatment as a general principle is limited.

On the other hand, modern bifurcation theory was mathematically formulated by Sattinger [10] and has since been employed in the fields of science and engineering. Bifurcation phenomena are one of the modes of failure for structures and materials. Examples include Euler buckling, lantern buckling of cylindrical shells, slip line formation in metallic materials, and the breakdown of sand, among other physical manifestations. Various methodologies concerning these phenomena are being developed currently. Moreover, bifurcation theory itself has evolved over time. The categorization of bifurcation points, including the classification of multiple bifurcation points with hill-top bifurcations, is gradually being organized, as seen in Koiter's initial imperfection theory and the catastrophe theory by Thompson & Hunt [11], both based on group theory. In the field of other science and engineering systems, the general principle of the description and utilization of symmetry has already been established. Symmetry is represented by symmetry manipulations consisting of rotational and mirror transformations. It is standard to use group theory [12], which expresses them mathematically, and is used to describe structures such as chemical applications [13] and crystal lattices [14]. Block diagonalization theory is established as a method of utilizing symmetry [15] such that the governing equations of a symmetric system can be decomposed into several independent equations by the coordinate transformation determined by the symmetry group of the system. This theory is systematically described in applied mathematics by group representation theory [16].

Recently, in the field of structural engineering and structural analysis, symmetries of structures have been described using the representation theory of groups, and studies on the utilization of symmetries have been carried out [17]- [40]. For example, Zloković [17,21] has applied block diagonalization to the undamped vibration problem for the first time, Golubitsky, Schaeffer [18] and adding Stewart [19] presented singularities and groups in bifurcation theory, Bossavit [20] exploited the symmetry for PDE problem, and Healy [22] and Tracy [23] also succeeded in applying block diagonalization to the undamped vibration problem of large-scale discrete systems with symmetry. Dinkevich [24–26] offered the block-diagonalization method, Murota and Ikeda [27] produced computational use of group theory in bifurcation analysis of symmetric structures, and [28] was a problem of random imperfections for structures of regular-polygonal symmetry. Ikeda and Murota [29,30] published bifurcation hierarchy of symmetric structures using block-diagonalization, and Ikeda and Ario et al. [31] presented block-diagonalization analysis of symmetric plates. Zingoni [32] worked the efficient computational scheme for the vibration analysis of high-tension cable nets and [33–35, 40] employed group-theoretical applications in solid and structural mechanics, symmetries and vibration problems such as natural frequencies. Zhang, Guest, Ohsaki [36] applied to symmetric prismatic tensegrity structures using duhedral group. Kaveh and Nikbakht [37] provided block diagonalization of Laplacian matrices of symmetric graphs via group theory and [38, 39] presented stability analysis of hyper symmetric skeletal structures using group theory.

The governing equations can be decomposed into multiple independent equations by transforming the (tangent) stiffness matrix of symmetric structures into block diagonals using appropriate coordinate transformations. Since these equations are independent of each other, they are highly suitable for parallel analysis such as parallel computers, and computational efficiency and reduction of arrangement capacity can be improved. The drastic reduction in the size of the matrix used for the numerical analysis by this theory not only reduces the operation cost but also improves the convergence stability of iterative calculations such as eigenvalue analysis.

However, in actually transforming the governing equations, the coordinate transformation matrix must be concretely made, and the problem remains in the consistency with the finite element method for various current elements. Also, large-scale complex eigenvalue analysis was required in order to carry out the natural vibration analysis of the system with damping, and this became a problem; a high computational cost was required. Currently, as a convenient modal analysis method for systems with damping, a method called *proportional damping*, in which the damping matrix is assumed to be proportional to the mass matrix and the stiffness matrix as well as an analysis method called *Rayleigh damping*, in which the off-diagonal term of the damping matrix is ignored and only the diagonal term is considered, are often used. However, both of these methods have the disadvantage that they can only be used for a limited damping matrix.

There are many hierarchic and various symmetries in physics or chemistry. The mathematic description of symmetry is well-known to use group theory. It has symmetry-breaking problem in nonlinear phenomena of periodic structure such as atom-array and nano-micro-structure on multi-scalable issue. In the computing engineering for solid and structural mechanics, BDM based on group theory used to study more efficient strategies for exploiting symmetry including the computing procedure. This topics is how simplified analysis for MFM problem and applied to its symmetry on the invariant.

This book aims to solve this problem by extending the trajectory to elements and nodal points. By using this concept, we propose a block diagonalization method that is consistent with the current finite element method [31]. For example, in the static problem, the block diagonalization method is extended to the nodal displacement including rotation, and in the dynamic problem, the matrices, etc., used in the stiffness, damping, mass, complex dynamic stiffness matrix, and control system are blocks diagonalized at the same time [47]. That is to say, in the structural analysis of the symmetric structure, the geometrical symmetry characteristic of the structure is mathematically expressed according to the physical law, and the utilization method of the block diagonalization is described as a mathematical analysis means and is designed to obtain a semi-analytical numerical solution based on the irreducible expression of the symmetry, in order to raise the computational efficiency by the speedup analysis of the parallel analysis. Previous studies are as follows.

1. Theory of block diagonalization and its applications for in-plane and out-of-plane linear displacement analysis of large discretized D_n invariant plates.

2. Comparison and analytical evaluation of the required analysis time results of the block diagonalization method and the conventional analysis method with the sequence capacity results are presented.

3. Extension of orbital concept for elements with finite symmetry and its development.

4. The block diagonalization for the symmetry to the conventional translational displacement is extended to the rotational displacement to clarify the mutual relationship between the symmetry of the translational displacement and the rotational displacement. And the assembly of the coordinate transformation matrix for the rotational displacement and the consistency of the finite element method based on the coordinate transformation matrix for the translational displacement.

5. D_n establishes a block-diagonalization method for dynamic-problems of invariant structures.

These theoretical studies approach the essence of the method of utilizing the symmetry of structural systems and are also applicable to general damping matrices. As an example of numerical calculation, the vibration problem of various axi-symmetric structural systems (the system which is identical to the dihedral group) is taken up, and the usefulness and versatility of the theory are verified.

We know well that we can make good use of the symmetry condition and reduce the degree of freedom of the problem in considering engineering problems such as structural mechanics and structural analysis. However, if the symmetry is mishandled, the solution may become inconsistent, which may move away from the reproducibility of the realization. It is known that even in simple cases, if the transformation laws are not consistent even in complex cases, the behavior of the deformation before and after the singularity of the nonlinear equation in a system of symmetric degrees of freedom differs from the prediction of the original symmetry condition (conjecture of linear deformation). As a method for mathematically describing symmetry operations and types of their structural systems, the "representation theory of groups" exists, and the effect demonstrates its power in conducting physical predictions. In this book, we would like to introduce the "block diagonalization method" and "block triangulation method," which the authors have developed in order to utilize representation theory for structural mechanics and the finite element method. This makes it possible to make concrete use of advanced symmetries such as ranks of symmetries and irreducible representations. In addition, we would like to introduce this calculation method as an effective means of the theoretical parallel analysis method by the transformation law "block diagonalization method" of this matrix.

1.2 Symmetric Structures

Symmetry is an important feature prevalent in nature and in artificial structures. From the periodic structure of crystals to the geometry of buildings, from the symmetry of molecules to the orbits of planets, symmetry plays a key role at all levels. However, despite the important role symmetry plays in structures, symmetry breaking may occur when structures experience large deformations or are subjected to external loads. In physics, symmetry breaking accompanied by critical phenomena is referred to as "spontaneous symmetry breaking" or simply "symmetry-breaking."

Symmetry is the transformation of a system from one state to another, and if the two states are equivalent, the system is said to be symmetric with respect to this transformation. In engineering, as in other fields, we can find structures that exhibit a certain degree of symmetry. Do you find a hidden symmetries shown in **Fig. 1.1**? For example, it would be recognized the several symmetries with mirror reflections of the dotted line shown in **Fig. 1.2** and symmetries of different scales exist in various aspects of nature and structure. The concept of symmetry not only simplifies the design of structures, but also brings harmony and beauty to them. Symmetry also plays a very important role in the buckling issues of structures. Symmetric and asymmetric errors in a system are usually a sign of a critical problem, implying that a problem such as a change in strength or a change in mode has occurred in the system. A symmetry is one of the strength tools for structural designs in engineering. However, it has its symmetry-breaking behavior in the post-buckling problem or nonlinear problem. Symmetry usually affects the stability of a structure. In the engineering field, a structure changes to a low symmetry or non-symmetric state after a break from a highly stable symmetric state, and this loss of symmetry is called the *symmetry-breaking* phenomenon. However, buckling analysis of high symmetry structures reveals that the structures exhibit a variety of different low symmetry patterns. This collapse process is a matter of bifurcation of the structure as it approaches a lower energy state. Since the bifurcation opens up many possibilities, an observer may assume that the outcome of the collapse is arbitrary.

There are various approaches for smart structures based on the units of scissor mechanism with foldable system. The shown in **Fig. 1.3** is a unique structure that is foldable and incorporates partial symmetric modules [50–52]. Similar to the bamboo basket discussed in Section 1.4, by repeating the same shape, it is possible to standardize the components, ensuring lightness,

FIGURE 1.1
Structures

storability, and structural durability. Figure **1.3**(a) is a cutting-edge bridge structure design that allows for smart bridging like a construction robot. This design, inspired by origami, was conceived at Hiroshima University and features a scissor-like deployment. Figure **1.3**(b), on the other hand, is the concept of foldable Truss-Z module (fTZM), which is designed so that the truss units are modularized to fold, and depending on how they are combined, the route of the bridge can be freely designed for movement [49]. In this way, the periodicity of structural units with folding mechanisms and standardized parts allows for the creation of various structures. Additionally, the required strength design can significantly deploy module structures as needed, extending beyond the field of space structures.

1.3   Origami and Symmetry

The strength of an origami largely depends on its structure and design, especially symmetry. How symmetry affects strength can be considered in terms

FIGURE 1.2
Symmetry of structures

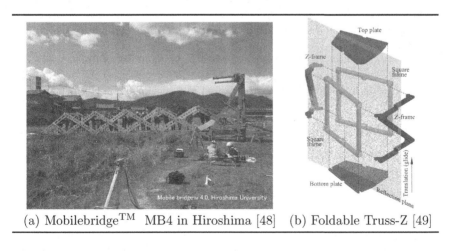

(a) Mobilebridge™ MB4 in Hiroshima [48]   (b) Foldable Truss-Z [49]

FIGURE 1.3
Symmetry and deployable structure

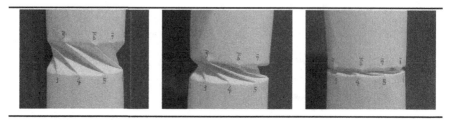

FIGURE 1.4
Foldable cylinder based on twist buckling of origami. (Refer Int. J. of Non-Linear Mech. 40 (2005) [41].)

of the following:

- Symmetric origami structures tend to distribute forces evenly. For example, if the folds of an origami are evenly spaced, the structure can distribute the load evenly throughout, resulting in increased strength. This principle is also frequently used in structural engineering and architecture.

- The placement of the folds has a great effect on the strength of the origami. If the folds are symmetrically placed, the force will be evenly distributed and the origami shape will be stable. On the other hand, if the folds are arranged asymmetrically, there is a possibility that the force will be concentrated in one part, and the shape will become unstable.

- Symmetrical origami designs are typically complex, and this complexity can add strength. For example, the origami pattern called *Miura-ori* is known for its amazing strength and flexibility due to its unique folding and symmetry.

However, although symmetry affects strength, more symmetry does not necessarily mean more strength. Many other factors, such as the design, materials used, and folding technique, also affect the overall strength of the origami.

1.3.1   Relationship between Kresling Folds and Symmetry

Kresling pattern or *Kresling fold* is a hot topic at the intersection of origami and structural mechanics. This pattern is found in nature and has a very efficient stretch. This pattern, which occurs naturally in bamboo, corn stalks, etc., was studied and named by Kresling [42]. The Kresling pattern is symmetrical because its basic structure is hexagonal. This symmetry allows uniform expansion and contraction of the structure, which allows for efficient motion and energy storage. The Kresling pattern is structurally superior. In particular, this origami structure has the ability to efficiently store and release energy through uniform expansion and contraction. The pattern is also strong and

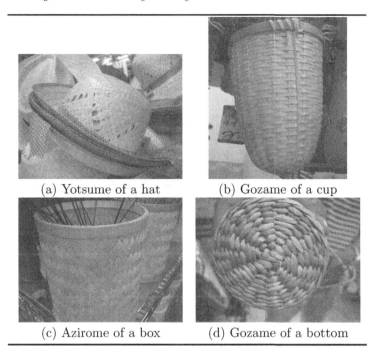

(a) Yotsume of a hat   (b) Gozame of a cup

(c) Azirome of a box   (d) Gozame of a bottom

FIGURE 1.5
Bamboo productions with periodic symmetry

lightweight and is being explored for use in a variety of applications, including architecture and robotics [41].

Kresling folds are resistant to buckling due to their unique structure. Buckling is a phenomenon in which a material suddenly deforms or breaks under compressive force. However, the Kresling pattern, due to its unique shape and symmetry, distributes forces evenly and prevents this buckling. This is similar to how bamboo, for example [44], resists the wind.

From the above points, the Kresling fold can be applied to symmetry and structural mechanics, and it has very interesting properties from the point of view of buckling.

1.4   Bamboo Knitting Patterns and Symmetry

In this section, there are things that make use of materials from the life of Japanese culture, for example, *bamboo knitting* using bamboo material shown in **Fig. 1.5**. Bamboo has been around for a long time, and has been used as a material that is strong, flexible, and easy to process (refer e.g. [43, 44]).

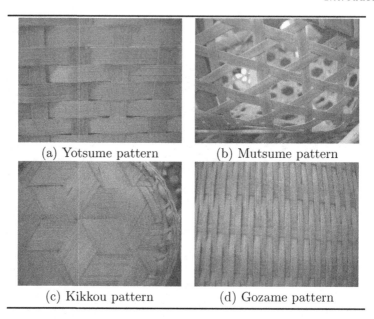

(a) Yotsume pattern   (b) Mutsume pattern

(c) Kikkou pattern   (d) Gozame pattern

FIGURE 1.6
Bamboo knitting patterns and symmetry

And there are various weaving methods depending on each bamboo basket or bamboo strainer to be woven.

Through several types of this bamboo weave, we would like to observe some symmetries in terms of the shape and structural strength of the bamboo weave. It is similar to the problem of laying tiles, but it is possible to weave bamboo strips into various structural shapes such as curves, lattices, flat/curved surfaces, and baskets. There is a traditional knitting pattern that utilizes the material, and its symmetry may be fascinated by the periodicity and beauty (artisticity) of the object. The beauty of its shape makes it a model for modern structures, and it has excellent features such as light weight, flexibility, functional design, and strength.

1.4.1   Basic Patterns of Bamboo Weaving and Their Symmetry

We introduce some common bamboo weaving patterns and their geometric symmetry. **Fig. 1.6** shows you how the patterns are made and what the symmetry of each pattern is. This introduction will deepen your understanding of the basic patterns and symmetry of bamboo weaving, giving you the knowledge to understand how bamboo weaving and symmetry are connected.

Main types of bamboo weaving: Gozame weaving, Yotsume weaving, Mutsume weaving, Ajiro weaving, Gozame weaving, and so on. The patterns and structures of bamboo weaving are diverse, and there is a wide range of types, from traditional techniques to modern designs. Here are some of them:

1. Yotsume Weave Pattern in **Fig. 1.6**(a): The Yotsume-ami (Four-Eye Weave) is one of the most basic methods in bamboo weaving. This technique involves crossing bamboo strips of the same width vertically and horizontally while maintaining equal gaps, forming a grid-like pattern. There are variations to this technique: a wider gap version is called *Renji-gumi*, and a narrower gap version is called *Ichimatsu-gumi*. Additionally, there is a slight variation called *Hishi Yotsume-ami*, where the bamboo strips are crossed at a slight angle, demonstrating that even within the same Yotsume-ami method, there can be subtle differences.

2. Mutsume Weave Pattern in **Fig. 1.6**(b): The Mutsume-ami (Six-Eye Weave), also known as Hexagonal weave or Basket weave, is a popular method in bamboo weaving due to its distinctive hexagonal pattern. This technique involves weaving six bamboo strips two diagonally and one horizontally to form a hexagonal pattern. Because the inclusion of diagonal weaves results in a robust structure, this method has been used for centuries in products that bear loads, such as bamboo colanders and baskets.

3. Kikkou Weave Pattern in **Fig. 1.6**(c): The Kikko-ami (Tortoiseshell Weave) is a weaving style named for its resemblance to the neatly tiled hexagonal pattern seen on a tortoise shell. This technique involves weaving six diamond shapes, highlighting a pattern of small hexagons as the work progresses.

4. Gozame Weave Pattern in **Fig. 1.6**(d): The Ajiro-ami (Ajiro weave) is a representative technique that uses wide, flat bamboo strips of the same width. The weave pattern is created by skipping every two or three crossings and offsetting the gaps. Because the vertical and horizontal weaves are knitted together without any gaps, the end result is an exceptionally sturdy bamboo weave. This method is versatile and can be adjusted to create unique variations, resulting in many different styles such as Ajiro-ami and variations. Consequently, it has not only been used in bamboo crafts but also in various household goods, such as ceilings and folding screens.

These patterns are flexible and can be made into a variety of shapes and designs. These are part of a bamboo weave pattern, each pattern chosen for its character, beauty, and use. The patterns and structures of bamboo weaving are infinitely diverse, and new designs and applications are constantly being explored.

1.4.2   Bamboo Weaving and Observations on Symmetry

The beauty and functionality of bamboo weaving arise from symmetry, Share common observations about how symmetry affects physical strength and stability. This will show the reader how this book will focus on the relationship between symmetry and bamboo weaving. Geometric symmetry refers to the property that an object remains unchanged through some kind of manipulation (rotation, reflection, translation, etc.). This section details the basic forms of geometric symmetry, including rotational symmetry, reflection symmetry, and translational symmetry. Geometric symmetry is closely related to the physical properties of an object, especially its strength and stability. This section explains how symmetry affects the strength of an object, with concrete examples. We explore how bamboo weaving patterns can be analyzed in terms of geometric symmetry. We specifically analyze the symmetry of various bamboo weaving patterns and detail what kind of symmetry operations each pattern can withstand. It may provide a basis for understanding the pattern (symmetry) of bamboo weaving and how it relates to its strength.

We explore how the physical properties of bamboo itself affect the strength of structures using it. Introduces the physical properties of bamboo, such as stretchability, durability, and elasticity, and explains how they influence the strength of bamboo braided structures. We will analyze in detail what kind of strength characteristics each bamboo weave pattern has. We can evaluate what kind of force the bamboo weave structure with various patterns is strong against and what kind of force it is weak against, and relate the result to geometric symmetry.

A theoretical consideration is given to how symmetry affects strength. We explain why patterns with high-symmetry are generally high in intensity, and elucidate the reasons from the point of view of geometry and physics. Through this chapter, the reader can understand how the bamboo weave pattern affects strength and what role symmetry plays in it. It consists of important basic knowledge for designing or evaluating bamboo weaving patterns.

1.4.3   Bamboo Braids Optimizing Symmetry and Strength

"Bamboo weaving for optimizing symmetry and strength" is a place where traditional craft techniques and modern scientific theories intersect, which can lead to new designs and structural improvements. This approach may open up new avenues for maximizing the strength, durability, and aesthetic appeal of bamboo weaves. Geometric symmetry determines the basic structure of bamboo weaving patterns. Highly symmetrical patterns are generally stronger and have the ability to distribute loads evenly. This is important to keep the structure strong and stable. Optimizing the strength of bamboo weaving patterns is important to increase the durability and functionality of bamboo products. Optimization of strength is achieved through bamboo weave pattern design, bamboo selection, and assembly methods. By optimizing symmetry and strength at the same time, the bamboo weave pattern can achieve both its functionality and aesthetic appeal. Increasing symmetry improves strength and creates beautiful patterns. By using the latest technology such as computer simulation and AI, optimization of symmetry and strength becomes more precise and efficient. These techniques offer great potential for designing new patterns and improving existing patterns. Bamboo is a renewable resource and bamboo weaving is a low environmental impact technique. Optimizing symmetry and strength improves the performance and longevity of woven bamboo products and increases their contribution to sustainability. bamboo weaving, which optimizes symmetry and strength, has the potential to create new value through the fusion of traditional techniques and modern science. These are all deeply connected when we talk about mathematics and/or physics such as bamboo weaving, the mystery of beauty and its relation to strength and universality. These elements are detailed below:

- Mathematical Elements: Bamboo weave patterns are defined by their inherently mathematical structure. These patterns are based on repetition, symmetry, and geometric regularity. These mathematical structures provide the foundation for bamboo weaving to achieve its strength and beauty at the same time.

- The Mystery of Beauty: While beauty is a subjective experience and highly dependent on individual perceptions, there are certain universal patterns of beauty. The beauty of bamboo weaving is enhanced by its complexity and simplicity, its order and chaos, and the mystery that arises from its inherent symmetry. These elements are closely related to geometric patterns to which the human visual system is naturally attracted.

- Relationship with Strength: The mystery of mathematical patterns and beauty is directly related to the strength of bamboo weaving. Symmetry and repeatability provide physical strength and structural stability. The bamboo weave pattern achieves high strength by evenly distributing the load and transferring forces effectively.

- Universality: The mathematical structure of bamboo weaving, its beauty and strength, have universality that transcends time and culture. Bamboo weave patterns have been valued for their functionality and beauty in various cultures and eras.

These patterns also have elements in common with structures and materials such as honeycombs, shells, lattices, crystals and periodic patterns in nature, demonstrating their Hierarchical and harmonious symmetrical universality.

1.5 Equilibrium and Symmetry using Balancing Toys, *Yajirobee*

In this section, let us consider the intuitive problem of balance using a static model of symmetry.

We can easily recognize that "structure is balanced" is equilibrium by the stational state of *Yajirobee* as shown in **Fig. 1.7**. Now, when the weights at both ends of the beam are the same, if they are equidistant from the center axis of the stirrer, they are symmetrical, the left and right moments are balanced at the center of the axis, and the beam is stationary. That is, for the balancing condition of the force of the target object,

- The sum of the decomposition forces in each direction is zero.

- The sum of the moments to be rotated is also zero.

Then, suppose that the wall is as shown in **Fig. 1.8**(a) instead of the arm on the left side of the stitcher, while maintaining this balancing condition, the right arm protrudes from the wall to be stationary. At this time, the left-right symmetry shape is no longer vanishing on the shape, but the moment of action due to the right-hand weight can be interpreted as balancing with the moment of resistance due to the wall. In other words, the center of the barrel

FIGURE 1.7
Yajirobee (やじろべえ) means balancing toys in Japanese

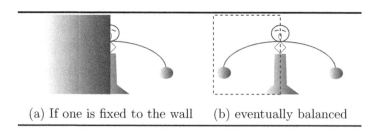

(a) If one is fixed to the wall   (b) eventually balanced

FIGURE 1.8
Hidden symmetry and force balance

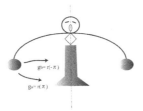

FIGURE 1.9
A symmetric transformation of the barrel.

does not rotate at this time, but the wall plays a role in the mirror symmetry condition that is perpendicular, and the left arm is hidden inside the wall as shown in **Fig. 1.8**(b). Since the restraint conditions of the deflection angle at the center of the beam are slightly different, in this case, information is lost for the dynamic rotation behavior due to the deflection (information missing due to the restriction of the symmetry condition). This symmetry condition is available based on the final stationary condition (statics), although it is a little different from the exact behavior of the filtration.

When **Fig. 1.7** is expanded to a three-dimensional space such as **Fig. 1.9**, Yajirobee swaps the left and right by rotating it clockwise or counterclockwise around the fulcrum. (It also maintains the symmetry that enables rotation conversion.)

2

Symmetry and Its Representation

Though symmetry widely exists, from natural micro elementary particles, atoms, molecules, and crystal structures to large-scale structures such as domes and shells of artificial structures, the research for examining the fundamental characteristics of the symmetry of the system with those symmetries is advanced mainly in the field of quantum mechanics and quantum chemistry, and utilization methods on the symmetry are being tried in each field. Group theory is now established as a method for mathematically expressing symmetry, and this book outlines a description method for utilizing this expression theory of symmetry for structural mechanics and computational mechanics as well as its basic idea [27, 30].

2.1 Basic Concept of a Group

This book describes the basic concept of a group for the mathematical representation of symmetry.

2.1.1 Group Definition

The product $g_i \cdot g_j$ of two operations $g_i, g_j \in G$ belongs to set G also [1]. Then, three conditions follow;

1. **Joint temperament:**For any three sources $g_a, g_b, g_c \in G$,

$$g_a(g_b \cdot g_c) = (g_a \cdot g_b)g_c \tag{2.1}$$

2. **Existence of the unit-source:** There is a special source $g(e)$ with the unit-source existence identity transformation e;

$$g(e) \cdot g_a = g_a \cdot g(e) = g_a, \quad {}^\forall g_a \in G \tag{2.2}$$

3. **Existence of inverse source:** There is an inverse for any original g_a of inverse existence;

$$g_a \cdot g_a^{-1} = g_a^{-1} \cdot g_a = g(e) \tag{2.3}$$

---

[1]Their operators and group products are called the source of G.

DOI: 10.1201/9781032670386-2

18

The set G is called *Group* when three conditions are satisfied. At this time, the original $g(e)$ is called the unit source. Moreover, in particular, the group in which the interchangeable $g_a \cdot g_b = g_b \cdot g_a$ for group multiplications holds is called a commutative group or *Abel group (Abelian group)*. For example, if we consider a set of n, $\exp(2k\pi i/n)$ $[k = 0, 1, \cdots, n-1]$ on a complex plane, it is obvious that this set meets all the requirements of the group. Thus, this set of n points forms a group. The original population of the group is called the order, and in this example the order is n.

2.1.2 Symmetry Operation

Symmetry operations describe the symmetry of nodes and elements. Here, the symmetry operation, for example, although the positions of various knot points are interchanged, as a result, the element and system is an operation such that it is present at an equivalent position, even if continuously repeated by combining the operation, the element and the system as a whole is a generic term of the transformation in the original equivalent position.

2.1.2.1 Definition of Symmetry Operation

Symmetry operations and symbols of symmetry groups shall conform to **Schönflies symbol**.

I, E, e: **identity:** The transformation is called an **identity transformation** when S is kept invariant, even if any transformation is made to the construct S.

i **inversion:** a coordinate component $S_i = (S_x, S_y, S_z)^{\mathrm{T}} \in$ S is

$$i : S_i \longrightarrow -S_i = (-S_x, -S_y, -S_z)^{\mathrm{T}} \tag{2.4}$$

The operation and the transformation to be performed shall be the inversion transformation.

C_n: **rotation:** Structure S with a clockwise rotation angle $R(2\pi/n)$ around any axis for the rotation operation, it is defined

$$C_n^j = R(2\pi j/n) \tag{2.5}$$

σ: **reflection:** The manipulation of a plane to give a pseudo representation of all points without destroying their original spatial arrangement. By continuing this transformation twice

$$\sigma^2 = e \tag{2.6}$$

It returns to the original position. In addition,

σ_h: **Mirroring with horizontal plane:** Conversion based on horizontal plane

$$\sigma_h : S_i \longrightarrow (S_x, S_y, -S_z)^{\mathrm{T}} \tag{2.7}$$

σ_v: **Mirror with vertical plane:** Conversion with vertical plane as base plane

$$\sigma_v : S_i \longrightarrow (-S_x, S_y, S_z)^{\mathrm{T}} \tag{2.8}$$

Or,

$$\sigma_v : S_i \longrightarrow (S_x, -S_y, S_z)^{\mathrm{T}} \tag{2.9}$$

σ_d: **Mirror with σ_d diagonal plane:** Transform based on diagonal plane

$$\sigma_d : S_i \longrightarrow (-S_x, -S_y, S_z)^{\mathrm{T}} \tag{2.10}$$

S_n: **improper rotation:** After $2\pi/n$ turns of the structure S around an axis, a horizontal mirror is used. It is defined

$$S_n^j = \sigma_h C_n^j \tag{2.11}$$

2.1.3   Systematic Classification for Symmetric Structures

This section describes how to classify any structure as S systematically and according to the type of group $(C_n, D_n, S_n, \cdots)$ (see **Fig. 2.1**). By systematically classifying them, it is easy to understand that different forms of structures retain the same physical properties (symmetry groups). Specific items are classified by stage as follows:

1. As a first step, it is classified into a linear structure, a structure with multiple higher-order axes, or other structures. The linear structure clearly belongs to the infinite group with the existence of the principal axis, and the multi-, higher- order axial structure belongs to the polyhedral group with high-symmetry.

2. Whether S has a group for operation $S_n(= \sigma C_n), n \geq 4$.

3. A group of structures is a C_s if S is not in the above group has no pivot and only a symmetric plane. If only the symmetry center is found, it is C_i, and if no other symmetry is found, it is C_1.

4. Whether or not there is C_2 axis that is perpendicular to the master axis that is the reference axis other than the above group. D_{nh} when this group is present and the mirror transformation that is satisfied for the horizontal plane, the mirror transformation only holds for some diagonal planes is classified as D_{nd} and others as D_n. Conversely, when there is no C_2 axis for the main axis, the mirror transformation is C_{nh} when the mirror transformation is established for the horizontal plane, C_{nv} when the is satisfied only for the diagonal plane, and the groups other than these two are called C_n.

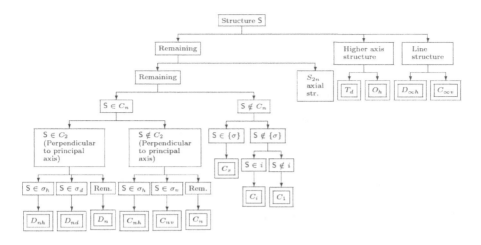

FIGURE 2.1
Systematic Classification Method for Symmetric Structures

2.2   Representation of Group

The group originally represents a collection of abstract "things" such as "symmetry" and "transformation". It seems to be unconstrained because we treat individual things as a general-purpose transformation rule. So, we need a means of making abstract things visible, called the *representation of the group*, and the representation theory of the group occupies the most important part of the group theory.

D-dimensional square matrix $D(g_i)$ exists for the original $g_1, g_2, g_3, \cdots$ of a group G and is a group. Relational expression of the product of the matrices corresponding to the original relational $g_i = g_j g_k$.

$$D(g_i) = D(g_j g_k) = D(g_j)D(g_k) \tag{2.12}$$

When this is true, the group D of matrices $\{D(g_1), D(g_2), \cdots\}$ is called the representation matrix of group G. For representation, D is often used, but this comes from the German Darstellung (representation). The dimension of the matrix is called the dimension of the representation. First, group requirement 1 is satisfied. We also satisfy requirement 2, because the product of matrices has a combination rule. For the identity E of G,

$$D(g_i) = D(Eg_i) = D(E)D(g_i) \tag{2.13}$$

Thus, $D(E)$ must be the unit matrix I. Since the unit matrix is identical in the set of matrices, the existence of identities is proved. In addition, from the relation to any source g_i and its inverse g_i^{-1}

$$D(g_i g_i^{-1}) = D(g_i)D(g_i^{-1}) = D(E) = \mathsf{I} \tag{2.14}$$

Since it is derived, $D(g_i^{-1})$ must be the inverse of $D(g_i)$. Since the inverse as a matrix also exists, the representation forms a group.

For the group consisting of the four complex numbers $\exp(in\pi/2)[n = 0, 1, 2, 3]$ listed in the previous section, the complex itself is a one-dimensional representation matrix. Next, we consider an example of a representation matrix where the group source is abstract, such as "a transformation operation," in the case of the translation group $\mathcal{T}$, which is listed in the previous section. Focusing on $\mathcal{T}_1^N = E$, the complex $z_n = \exp(i2\pi n/N)$ (where n is an integer) is the source $\mathcal{T}_1$. $\mathcal{T}_j$ is the corresponding z_n^j, and $D^{(n)} = \{z_n, z_n^2, z_n^3, \cdots, z_n^N (= 1)\}$ is a group of one-dimensional representation of $\mathcal{T}$. [2]

Generally, since different n give different representations, we regard n as the name of the representation. If $n = 0$, all elements of $D(0)$ are 1, so N sources correspond to one representation matrix. In the main, multiple groups correspond to a single representation matrix. If $n = 1$, all elements of $D(1)$ are different, and one representation matrix corresponds to one grouping source. Such a 1-to-1 correspondence is called a **faithful representation**. There is also a basis in the representation matrix such that there is a basis vector in the general matrix.

$$\mathcal{T}_j \psi_n(x) = \psi_n(\mathcal{T}_j^{-1}x) = \psi_n(x - ja) \tag{2.15}$$

In the case of the group consisting of transformations, there is a function. By writing the base function corresponding to the expression $D(n)$ as $\psi_n(x)$. From the fact that $\psi_n(x)$ is the basis of a one-dimensional representation. It means

$$\mathcal{T}_j \psi_n(x) = \exp\left(\frac{i2\pi nj}{N}\right) \psi_n(x). \tag{2.16}$$

Therefore, the following relationship is obtained

$$\psi_n(x - ja) = \exp\left(\frac{i2\pi nj}{N}\right) \psi_n(x). \tag{2.17}$$

This is called the Bloch theorem that holds for the wave function of the grain in the periodic potential. By introducing the function $u_k(x)$ in $\psi_n(x) = \psi_k(x) = \exp(-ikx)u_k(x)$, $k = 2\pi n/Na$ a using k instead of n,

$$\psi_k(x - ja) = \exp(ikja)\psi_k(x) \tag{2.18}$$

$$u_k(x - ja) = u_k(x) \tag{2.19}$$

---

[2]Two cyclic groups have an **irreducible representation** of one dimension.

TABLE 2.1
Index of D_4

χ^μ	$\{e\}$	$\{r^2\}$	$\{r, r^3\}$	$\{s, sr^2\}$	$\{sr, sr^3\}$
$\chi^{(1,1)}$	1	1	1	1	1
$\chi^{(1,2)}$	1	1	1	-1	-1
$\chi^{(1,3)}$	1	1	-1	1	-1
$\chi^{(1,4)}$	1	1	-1	-1	1
$\chi^{(2,1)\pm}$	2	-2	0	0	0

While $\psi_k(x)$ is called a Blotcher function, which is represented by the product of a plane wave and a periodic function. Bloch's theorem is easily extended to three-dimensional crystals. Quantum mechanically, the *Hamiltonian* is kept constant in the transformed T_j.

Next, we consider a slightly more complicated expression. The group of four complex numbers listed first is the same as a collection of four points on the x-y plane. Now, consider a transformation that moves between four points by rotation about the z-axis. For convenience, the four points are named A(1,0), B(0,1), C(-1,0) and D(0,-1). We will refer to the rotation of R_1, R_2, R_3, and R_4 around the z-axes $90°$, $180°$, $270°$, $360°$, respectively, as R_1, R_2, R_3, R_4. This set of four points in four different rotations is invariant as a set simply by shifting from one another.

To ascertain whether or not to form a group, it is convenient to make a "**group table**." This is based on the vertical and horizontal orientations, and the product (vertical × horizontal) is written at the intersection position. For the four rotations that are currently being handled, the result is as shown in **Table 2.1**.

The tables show that R_4 is the source of identity, that the product is commutative, and that R_1 and R_3 are reversed from each other. Needless to say, the combination rule of product. This group is usually called C_4. It is also evident from the properties of rotation that the group C_4 is a cyclic group of order 4. Since there is a one-dimensional representation because it is a cyclic group [3] from a slightly different point of view, consider the representation of C_4. Generally, when the point (x, y) moves to (x', y') due to the rotation of the angle alpha around the z-axis, the two points are related by the matrix as follows.

$$\begin{pmatrix} x' \\ y' \end{pmatrix} = \begin{pmatrix} \cos\alpha & -\sin\alpha \\ \sin\alpha & \cos\alpha \end{pmatrix} \begin{pmatrix} x \\ y \end{pmatrix} \tag{2.20}$$

If the transformation matrix is represented as $R(\alpha)$, then $R(\pi/2), R(\pi), R(3\pi/2)$, and $R(2\pi)$ are R_1, R_2, R_3, and R_4 representation matrices, respectively. The

[3]The four complex numbers listed above represent the four rotation operations.

collection of four matrices is a two-dimensional representation of group C_4. The concrete form of each matrix is as follows.

$$R\left(\frac{\pi}{2}\right) = \begin{pmatrix} 0 & -1 \\ 1 & 0 \end{pmatrix}, \quad R(\pi) = \begin{pmatrix} -1 & 0 \\ 0 & -1 \end{pmatrix} \tag{2.21}$$

$$R\left(\frac{3\pi}{2}\right) = \begin{pmatrix} 0 & 1 \\ -1 & 0 \end{pmatrix}, \quad R(2\pi) = \begin{pmatrix} 1 & 0 \\ 0 & 1 \end{pmatrix} \tag{2.22}$$

Several representations are generally possible for a single group. In the previous one-dimensional representation example, we showed that there are various kinds of representations of the same dimension, but what is the relationship between the different representations of dimensions?

2.3   Simplification of Expression

Some representations of group C_4 differed in dimensions. In general, all representations belonging to the different representations D, D' of the group G are by the appropriate regular matrix S (determinant non-zero square matrix: there is an inverse matrix).

$$D'(a_i) = S^{-1}D(a_i)S, \quad a_i \in G \tag{2.23}$$

When they are tied together, the two representations are said to be equivalent. Representations with different dimensions are non-equivalent. Looking closely at the matrices in Eqs. (2.26) and (2.27), it sets to the following;

$$R\left(\frac{3\pi}{2}\right) = -R\left(\frac{\pi}{2}\right) \tag{2.24}$$

$$R(2\pi) = -R(\pi) = I. \tag{2.25}$$

Thus, we find the regular matrix S, where $R(\pi/2)$ is diagonalized by $S^{-1}R(\pi/2)S$. This is accomplished by arranging the $R(\pi/2)$ normalized eigenvector $\boldsymbol{x}_1, \boldsymbol{x}_2$, so that the four complex numbers listed above represent the four rotation operations. Assuming $S = (\boldsymbol{x}_1, \boldsymbol{x}_2)$ which are vertical vectors, the inverse matrix is derived from the relation $(S^{-1})_{kl} = S_{lk}^*$, where $*$ represents a complex conjugate.

$$S = \begin{pmatrix} \frac{1}{\sqrt{2}} & \frac{1}{\sqrt{2}} \\ \frac{i}{\sqrt{2}} & \frac{-i}{\sqrt{2}} \end{pmatrix}, \quad S^{-1} = \begin{pmatrix} \frac{1}{\sqrt{2}} & \frac{-i}{\sqrt{2}} \\ \frac{1}{\sqrt{2}} & \frac{i}{\sqrt{2}} \end{pmatrix} \tag{2.26}$$

The result of the diagonalization of $R(\pi/2)$ becomes

$$S^{-1}R\left(\frac{\pi}{2}\right)S = \begin{pmatrix} i & 0 \\ 0 & -i \end{pmatrix} \tag{2.27}$$

At this time, the diagonal component becomes an intrinsic value of $R(\pi/2)$. The other three matrices can be displayed diagonally using the same S.

$$S^{-1}R\left(\frac{3\pi}{2}\right)S = \begin{pmatrix} -i & 0 \\ 0 & i \end{pmatrix} \tag{2.28}$$

$$S^{-1}R\left(\pi\right)S = \begin{pmatrix} -1 & 0 \\ 0 & -1 \end{pmatrix} \tag{2.29}$$

$$S^{-1}R\left(2\pi\right)S = \begin{pmatrix} 1 & 0 \\ 0 & 1 \end{pmatrix} \tag{2.30}$$

This new four-diagonal two-dimensional representation is equivalent to the original two-dimensional representation. It is easy to see that a set of only 1-1 components of a diagonal matrix is a one-dimensional representation of C_4. Similarly, a set consisting of only 2-2 components has a one-dimensional representation. Therefore, the new 2D representation uses two 1D representations in the following;

$$D'(R_k) = \begin{pmatrix} D^{(1)}(R_k) & 0 \\ 0 & D^{(2)}(R_k) \end{pmatrix}, \quad k = 1, \cdots, 4 \tag{2.31}$$

When a specific representation is generally represented by a representation of a lower dimension arranged diagonally in several diagonals, the former is said to be "the direct sum" of the latter. The former dimension is given by the sum of the latter dimensions. The transformation of a representation into a direct sum of the representations of a lower dimension by an equivalent transformation is called an *abbreviation*. Expressions that can be reduced are called **reducible expressions**, and expressions that cannot be reduced are called **irreducible expressions**. In the above example, the two-dimensional representation $\{R(\pi/2), R(\pi), R(3\pi/2), R(2\pi)\}$ of C_4 is converted to the direct sum of the two irreducible one-dimensional representations by the regular matrix S of Eq.(2.31). It is converted to a direct sum of two irreducible one-dimensional representations. There is a method that uses the table of the index of irreducible expression how the expression is reduced. An indicator is the sum of the diagonal components of a representation matrix. For example, representation D is reduced to the form of the sum of irreducible representations, as follows;

$$D = \sum_i D^{(i)} \tag{2.32}$$

Since the sum of the diagonal elements is immutable according to the simplification procedure (2.28)-(2.30), the index $\chi(g_k)$ of the representation D for a given original g_i is expressed as follows;

$$\begin{aligned} \chi(g_k) &= \sum_i \chi^{(i)}(g_k) \\ &= \sum_\lambda c_\lambda \chi^{(\lambda)}(g_k) \end{aligned} \tag{2.33}$$

TABLE 2.2

The table of the character D_6

χ^μ	e	$2C_6$	$2C_3$	C_2	$3C_2$	$3C_2''$
$\chi^{(1,1)}$	1	1	1	1	1	1
$\chi^{(1,2)}$	1	1	1	1	−1	−1
$\chi^{(1,3)}$	1	−1	1	−1	1	−1
$\chi^{(1,4)}$	1	−1	1	−1	−1	1
$\chi^{(2,1)\pm}$	2	1	−1	−2	0	0
$\chi^{(2,2)\pm}$	2	−1	−1	2	0	0

Where c_λ represents the number of occurrences of the same abbreviation. This relationship must hold for all the origins belonging to the group currently being addressed. Therefore, we can know how it is reduced by calculating the index of a certain expression over all the origins and comparing it with the table of the index of irreducible expression which may appear when the expression is reduced. **Table 2.2** can be viewed as a table of indicators, so superimposing the indicators for the four elements from Eqs. (2.26) and (2.27) satisfies the relationship of Eq.(2.33). Even if other expressions are mixed in the table of indicators, we can know the result of the reduction by considering what combination there must be for Eq.(2.33) to be satisfied with all the origins. The reduction of representation matrices is significant for practical use when we have to deal with representations of large dimensions in complex problems.

2.4   Supplementary Information on Group Tables

In Section 2.2 of this book, we describe that group tables are useful for organizing groups, but when we focus on a column or a row of group tables, the groups always appear only once and never twice. This property is called the recombination theorem and is proved as follows. Using one element g of a group $G = \{g_1, g_2, \cdots, g_n\}$, we make a set of $\{gg_i\}(i = 1, \cdots, n)$ or $\{g_i g\}$. From the definition of a group table, this procedure results in one row of the group table. For example, fetching two elements from one row

$$gg_i = gg_j, \quad g_i \neq g_j \tag{2.34}$$

In this case, multiplying both sides by g^{-1} results in $g_i = g_j$, contrary to the assumption. It must therefore be $gg_i \neq gg_j$. Since gg_i and so on can be any of the group sources from the definition of the group, the source of the group always appears once in one row or column of the group table, as mentioned earlier.

2.5 Subgroups and Conjugate Elements

When the subset G_{sub} of group G forms a group with the same meaning as G, G_{sub} is called a subset of G. The identity only satisfies the condition of the subgroup, and G itself is also a subgroup of itself, but since the two are trivial subgroups, we call them **true subgroups** to distinguish them if there are other subgroups. If the element of G_{sub} is represented as $\{b_1, b_2, \cdots\}$, in order for G_{sub} to form a group, $b_i b_j$ must be included in G_{sub} and the inverse element of each element must be included in G_{sub}. These two, and the fact that G_{sub} is a subset of G, verify that G_{sub} satisfies the group's four prerequisites. Two elements of a group, g_i and g_j, belong to the same group, depending on one element g and its inverse.

$$g_j = g g_i g^{-1} \tag{2.35}$$

When it is related, g_i and g_j are said to be conjugate. If you choose either identity or g_i itself as g, you can see that g_i is coupled to itself. All sets of elements that are conjugated to one source are called **conjugates**, or **merely classes**merely classes. If g_i and g_j are conjugated, and g_i and g_k are conjugated, then g_j and g_k are conjugated to each other, so all two elements of the class are conjugated to each other. Here, we describe the relationship between subgroups and classes. Subgroup $G_{\text{sub}} = \{b_1, b_2, \cdots\}$ is called the *conjugate subgroup* of G_{sub} and the set $g G_{\text{sub}} g^{-1}$ containing the original group of G, the original a, and the inverse. It is clear that $g G_{\text{sub}} g^{-1}$ is a subset of the group G, and the fact that it meets the requirements of the group can also be shown by G_{sub} being a group. (It is only necessary to show that the accumulation of elements is included in the set along with the inverse.) If we use the recombination theorem twice, we can show that $g b_i g^{-1} \neq g b_j g^{-1}$ when $b_i \neq b_j$, so the elements of G_{sub} and $g G_{\text{sub}} g^{-1}$ have a 1-to-1 correspondence. When there is a group source 1-to-1 correspondence, the two groups are said to be of the same type. $g G_{\text{sub}} g^{-1}$ and G_{sub} are likely to differ because the subgroups do not generally contain all the elements that are conjugate to each other. For any original G_{sub} of G in special cases

$$g G_{\text{sub}} g^{-1} = G_{\text{sub}} \tag{2.36}$$

There may be the subgroup G_{sub} that satisfies. Such subgroups are called invariant subgroups (or normal subgroups). The invariant subgroups are subgroups as well as classes. The subgroups of the commutative groups (Abelian groups) are always invariant subgroups. For the group consisting of four points

$\exp(in\pi/2)[n = 0, 1, 2, 3]$ on the complex plane listed in Section 2.2, it is easily ascertained that the subset consisting of 1 and $\exp(i\pi)$ is an invariant subset.

2.5.1   Isomorphism and Quasi-isomorphism

Consider the relationship between two or more groups. There are two arbitrary groups G, G', and there is a 1-to-1 correspondence between the original $g \in G$ and $g' \in G'$, product operation $g_i \, g_j = g_k \in G$ and $g'_i \, g'_j = g'_k \in G'$ holds, the groups G and G' are said to be **isomorphic** and are written as $G \simeq G'$. A mapping from group G to group G', $f : G \longrightarrow G'$ for any original $g_a, g_b \in G$ of group G,

$$f(g_a \circ g_b) = f(g_a) \diamond f(g_b) \tag{2.37}$$

When it is satisfied, f is called G to G' homomorphic mapping (homomorphism). Here, $\circ$ denotes the operation of G, and the operation of G' is denoted by $\diamond$. In this case, the image of unit e of G is the source e' of G', and the original g_a^{-1} image is the inverse of g_a image.

$$f(e) = e', \quad f(g_a^{-1}) = [f(g_a)]^{-1}. \tag{2.38}$$

2.5.2   Conjugate Elements and Types

Another original g_j of the same transformed the original g_i of the group G group, and the original of $g_j \, g_i \, g_j^{-1}$ form was conjugated. All the original sets that are conjugated to the original g_i are called g_i conjugates (conjugacy class). For example, Dihedral group C_{nv} is defined as the following;

$$C_{nv} \equiv \{1, r, \cdots, r^{n-1}, s, sr, \cdots, sr^{n-1}\} \tag{2.39}$$

(For detail, refer to Section 5.1.) Where r is the rotation transformation, s is the mirror transformation, and 1 is the identity transformation. Incidentally, since it is customary to denote the Dihedral group C_{nv} as D_n [27, 30], the latter symbol will be used in this book. There is a $r^n = s^2 = 1$ relationship between them, and the conjugate relationship by s is $srs^{-1} = srs = r^{-1} = r^{n-1}$ and it has one conjugate of $sr^{-1}s^{-1} = (srs^{-1})^{-1} = r$. That is, $\{r, r^{n-1}\}, \{r^2, r^{n-2}\}, \cdots$, where n is odd, there are sums of $(n+3)/2$, and n is even, there are $(n+6)/2$ conjugates.

2.5.3   Subgroups and Immutable Subgroups

When the subset G_{sub} of group G constructs a group of G operations, G_{sub} is called a subset of G (sub-group) and is written as $G_{\text{sub}} \subseteq G$. For example, a group source g_a, g_b

$$g_a, g_b \in G_{\text{sub}} \Longrightarrow g_a, g_b \in G_{\text{sub}} \tag{2.40}$$

$$g_a \in G_{\text{sub}} \Longrightarrow g_a^{-1} \in G_{\text{sub}} \tag{2.41}$$

Necessary and sufficient conditions must be satisfied. Also, subgroup G_{sub} is an additional requirement.

$$g_c \in G \implies g_c G_{\text{sub}} g_c^{-1} = G_{\text{sub}} \qquad (2.42)$$

When it is satisfied, G_{sub} is called an *invariant subset* of G (*invariant subgroup*).

2.5.4 Translation Group and Rotation Group

In this section, one-dimensional crystals are considered, and the lattice constant is a. In this crystal, a physical quantity is given as a function of the coordinate x as $f(x)$. From the periodicity of the crystal, it is considered that this function does not change by shifting x by an integer multiple of a. This is expressed as

$$f(x + ak) = f(x), \quad k \in \mathbf{Z} \qquad (2.43)$$

Let us say that this x shifted by $x + ak$ with the symbol $\mathcal{T}_k$, $\mathcal{T} = \{\mathcal{T}_k; k = 0, \pm 1, \pm 2, \cdots \}$ The set also is satisfied all the requirements of the group. This is similarly defined for two-and three-dimensional crystals, and this assembly of translational operations is called **a translational group**.

For the sake of simplicity, we will limit this translational group T to one dimension and look a little more closely at the nature of this translational group $\mathcal{T}$. Although crystals are invariant in translational operations, real crystals are not strictly invariant because they are finite in size and have edges. However, if we suppose a sufficiently large crystal, the contribution of the edge is very small, and we need hardly consider it. A periodic boundary condition is used to mathematically remove the edge effect. In the case of one dimension, it corresponds to making chainlike crystals into rings. If the length of the ring is $L = Na$ and N is large enough, the bending of the chain can be ignored and the treatment as a one-dimensional system is justified.

The original number of groups $\{\mathcal{T}_n\}$ is N from $n = 0$ to $n = N - 1$. That is, this translation group is a group of orders N. Repeating $\mathcal{T}_1$ m times gives $\mathcal{T}_m$, and because of the periodicity of the ring.

$$(\mathcal{T}_1)^N = \mathcal{T}_0 = E \qquad (2.44)$$

Notice that all the origins of this translation group are represented by the power of $\mathcal{T}_1$. A group with one source of power is called a **cyclic group**. Any origin of the cyclic group multiplied by the position is the identity. And the product between the original of the cyclic groups is commutative. In the example above, commutation is considered.

$$\mathcal{T}_i \mathcal{T}_j = \mathcal{T}_1^i \mathcal{T}_1^j = \mathcal{T}_1^{i+j} \qquad (2.45)$$

A group that is trivial from Eq.(2.45) and whose product is commutative is generally called a "commutative group (Abelian group)."

2.6   Group Representation

To investigate the structure of the group, it is necessary to "express" the *operation* quantitatively. Generally, a group considers a vector space and expresses the group as a first-order transformation in it. This is called *group representation*, and group representation theory occupies the most important part of group theory.

2.6.1   Representation Matrix of Group G

Assume that the matrix representation $T(g_i)$ is given for the source $g_i(i = 1, \cdots, |G|)$. [4]

The relationship between any two group sources is the following;

$$g_i \, g_j = g_k, \quad {}^\forall g_i, g_j \in G \tag{2.46}$$

and, its matrices are related as follows;

$$T(g_i)T(g_j) = T(g_k), \quad g_i, g_j \in G \tag{2.47}$$

When it holds, the set of matrices $T(g_1), \cdots, T(g_n)$ is called the (matrix) representation of group G. [5]

In addition, the basic transformation for the representation matrix is the unit matrix for the original unit $g(e)$ and inversely

$$\begin{aligned} T(g(e)) &= T(g) \cdot T(g^{-1}) \\ &= T(g) \cdot [T(g)]^{-1} \tag{2.48} \\ &= \mathrm{I} \tag{2.49} \end{aligned}$$

Generally, the relationship between the group source and the representation matrix is not always 1-to-1, but n-to-1.

2.6.2   Representation Theory in General

In this section, we describe a theorem that generally holds for representation matrices. The following theorem deals with the finite group G of the order $|G|$ of a group. [6] The finite group representation does not lose generality even if only the *unitary representation* is considered by the equivalence transformation. Therefore, in this book, we consider only unitary expressions for the following expressions.

---

[4] Generally, $T(g)$ is called the representation matrix (represent matrix) and is used hereafter.

[5] Equation (2.47) is also called homomorphic mapping condition and homomorphic mapping $T: g \to T(g)$.

[6] Note that the same conclusions are correctly derived for groups called **compact groups** even in the infinite groups.

2.6.2.1 Schur's Lemma 1

The following two Schur's lemma have the crucial nature of the representation matrix presented by the irreducible matrix.

$$T^\mu(g)H = HT^\nu(g), \quad g \in G \tag{2.50}$$

Here, let G consist of complex matrices.
    The matrix H in which this equation (2.50) follows

1. $H = O$, or,

2. Is the square matrix $\det |H| \neq 0$

For 2. above, there is an inverse matrix H^{-1} of H,

$$H^{-1}T^\mu(g)H = T^\nu(g), \quad g \in G \tag{2.51}$$

It can thereby be obtained and the canonical representation $T^\mu(g), T^\nu(g)$ is equivalent. If $T^\mu(g), T^\nu(g)$ is not an equivalency, then H does not exist to satisfy the expression (2.51). Thus, in this case, the matrix H satisfying Eq.(2.50) is limited to a zero matrix.

2.6.2.2 Schur's Lemma 2

For all representation matrices $T(g)$ of the representation T of group G

$$T(g)H = HT(g), \quad g \in G \tag{2.52}$$

The matrix H satisfies the unit matrix I multiplied by the complex number c if the representation $T(g)$ is irreducible.

$$H = cI \tag{2.53}$$

2.6.3 Reducible and Irreducible Representations

An equivalence relation by a regular matrix U with two representations of a group T, T' representation matrices $(T(g), T'(g) \in G)$

$$T'(g) = U^{-1}T(g)U \tag{2.54}$$

When it holds, the representation matrix $T'(g)$ is

$$T'(g) = \begin{pmatrix} T_1(g) & O \\ O & T_2(g) \end{pmatrix} \tag{2.55}$$

It can then be transformed into a block diagonal matrix. The representation T is said to be an approximate representation (reducible representation), for example, the representation that T_1, T_2 can no longer make is called an abbreviated representation (irreducible representation).

T_1, T_2 in Eq.(2.55) also represents the group G, and the representation T is the direct sum of T_1 and T_2. Representation of it becomes possible.

$$T = T_1 \bigoplus T_2 \tag{2.56}$$

This means that the dimension of the representation T is equal to the sum of the dimensions of T_1 and T_2. Thus, a conservative representation can be decomposed by some transformation into a direct sum of several irreducible representations.

2.6.4   Indicators and Their Properties

The sum of the diagonal components of the square matrix is called the index (character) of the matrix;

$$\chi = \mathrm{Tr}\, Q = \mathrm{Tr} \begin{pmatrix} Q_{11} & \cdots & Q_{1N} \\ \vdots & \ddots & \vdots \\ Q_{N1} & \cdots & Q_{NN} \end{pmatrix} = \sum_{i=1}^{N} Q_{ii} \tag{2.57}$$

Where Tr is an operator that adds the diagonal components of the matrix. This distinguishes between irreducible and irreducible representations of a given representation matrix and is itself a representation matrix. This is when the irreducible representation matrix $T(g)$ involves the group element g;

$$\chi(g) = \mathrm{Tr}\, T(g) = \sum_{i=1}^{N} T_{ii}(g) \tag{2.58}$$

The character of these matrices is invariant under the similarity transformation. In addition, there is a table of the character showing the relation between the type of irreducible expression and the character. For example, **Tables 2.1** and **2.2** show the index tables of Dihedral groups D_4 and D_6 when the symbolic method of irreducible expression is followed.

See the literature [13, 54, 55] for details.

2.6.5   Orthogonal Irreducible Representation Matrix

We define the irreducible representation matrix of the group G as

$$T^{\mu}(g) = T_i^{\mu}(g), \quad i = 1, \cdots, a^{\mu}, \quad g \in G, \quad \mu \in R(G) \tag{2.59}$$

Here, μ represents an arbitrary irreducible representation of the group G, and $R(G)$ represents the entire irreducible representation. Where μ denotes any irreducible representation of group G and $R(G)$ denotes the entire irreducible representation, respectively.

Also, the matrix components $T_{ij}^{\mu} \unrhd T_{k\ell}^{\nu}$ of the two irreducible representations μ and ν with the group G. The **irreducible representation orthogonality theorem** follows [13]

$$\sum_{g \in G} T_{ij}^{\mu}(g) T_{k\ell}^{\nu*}(g) = \frac{|G|}{\sqrt{d_\mu d_\nu}} \delta_{ik} \delta_{j\ell} \delta_{\mu\nu}, \quad \mu, \nu \in R(G) \tag{2.60}$$

Where d_μ and d_ν are the orders of each irreducible representation, $|G|$ is the number of elements g in the group, and δ_{ab} is the Kronecker. By this theorem, the equation of motion with the same transformation (3.6) in the next chapter is decomposed into independent equations for each irreducible representation. The representation matrix $T(g)$ of each structure is transposed as

$$H^{\mathrm{T}} T(g) H = \bigoplus_{\mu \in R(G)} \bigoplus_{i=1}^{a^{\mu}} T_i^{\mu}(g), \quad {}^{\forall} g \in G \tag{2.61}$$

Finding the transformation matrix H for block diagonalization is an individual theory that is different for each structural example. Also, a^{μ} is the multiplicity of the irreducible representation μ in the representation matrix $T(g)$, and is expressed as;

$$a^{\mu} = \frac{1}{|G|} \sum_{g \in G} \chi(g) \overline{\chi^{\mu}(g)}, \quad \mu \in R(G) \tag{2.62}$$

In the formula, $\overline{\cdots}$ represents the complex conjugate, $\chi(g)$ is the representation matrix $T(g)$, and $\chi^{\mu}(g)$ is $T^{\mu}(g)$. This means the sum (index) of the diagonal terms of $\mu(g)$. A general principle is constructed for this irreducible representation matrix $T^{\mu}(g)$, and the transformation matrix H between the representation matrix $T(g)$ of each structure and the irreducible representation matrix is obtained. Therefore, the basic idea of the *representation theory of groups* is to describe the symmetry of each structure. This theorem is proved by Schur's lemma. When we describe the theorem as mentioned earlier in a somewhat more embodied form, we can draw five conclusions:

1. Suppose that the possible irreducible representations for one group are $|G|$ varieties $T^{\mu}(g)$. The sum of the squares of the degree d_μ of the representation matrix of the irreducible representation μ over all irreducible representations is equal to the number of symmetry operations in the group.

$$\sum_{\mu \in R(G)} d_\mu^2 = |G| \tag{2.63}$$

2. For a certain irreducible representation $T^{\mu}(g)$, let $\chi^{\mu}(g)$ be the index of each representation matrix corresponding to the symmetry

operation. The squared addition for all symmetry operations is also equal to the number of symmetry operations in the group.

$$\sum_{g \in G} [\chi^\mu(g)]^2 = |G|, \quad \mu \in R(G) \tag{2.64}$$

3. Create an index of each representation matrix corresponding to the same symmetric operation g for the two default representations $T^\mu(g), T^\nu(g)$ of the two groups. If it is $\chi^\mu(g), \chi^\nu(g)$, then the product plus all symmetry operations is equal to zero.

$$\sum_{g \in G} \chi^\mu(g)\chi^\nu(g) = 0, \quad \mu \neq \nu, \quad \mu, \nu \in R(G) \tag{2.65}$$

From Eq.(2.60) for example, let $i = j, k = \ell$ be a diagonal component.

$$\sum_{g \in G} T^\mu_{ii}(g)T^\nu_{kk}(g) = 0, \quad \mu \neq \nu \tag{2.66}$$

Taking the sum of i and k on the left side, it is easily leaded the following:

$$
\begin{aligned}
\sum_i \sum_k \sum_{g \in G} T^\mu_{ii}(g)T^\nu_{kk}(g) &= \sum_{g \in G} \left(\sum_i T^\mu_{ii}(g) \right) \left(\sum_k T^\nu_{kk}(g) \right) \\
&= \sum_{g \in G} \chi^\mu(g)\chi^\nu(g) = 0
\end{aligned} \tag{2.67}
$$

4. Within one representation, the indices of the representation matrices of symmetry operations belonging to the same class are equal.

5. The number of irreducible representations possible in a group is equal to the number of such classes.

2.6.6 Linear Transformation Group

The linear operator T on the vector space V^N is represented by the matrix;

$$T = \begin{pmatrix} T_{11} & \cdots & T_{1N} \\ \vdots & \ddots & \vdots \\ T_{N1} & \cdots & T_{NN} \end{pmatrix} \tag{2.68}$$

In particular, in $\det |T| \neq 0$, the inverse matrix T^{-1} exists, and even in the set of the whole matrix, when the inverse element and the identity element exist, they form a group. A group of matrices whose matrix elements are complex numbers is called a complex general linear transformation group.

$$GL(N, C) \equiv \{T_{ij}|, i, j = 1, \cdots, N; \ T_{ij} \in C, \ \det |T| \neq 0\} \tag{2.69}$$

In the case of a real number, it is called a real general linear transformation group of order n. Also, a special subgroup of a matrix in which both groups $\mathrm{GL}(N, \boldsymbol{C}), \mathrm{GL}(N, \boldsymbol{R})$ satisfy $\det |T| = 1$. It is defined as

$$\mathrm{SL}(N, \boldsymbol{C}) \equiv \{T \in \mathrm{GL}(N, \boldsymbol{C}) | \det |T| = 1\} \tag{2.70}$$

Furthermore, the whole group of unitary matrices consisting of $N \times N$ is called a **unitary group**.

$$\mathrm{GU}(N, \boldsymbol{C}) \equiv \{U | U_{ij} \in \boldsymbol{C}, U^* U = \mathrm{I}\} \tag{2.71}$$

In particular, the group formed by the unitary matrix U that satisfies $\det U = 1$ is called the **special unitary group**.

$$\mathrm{SU}(N, \boldsymbol{C}) \equiv \{U \in \mathrm{GU}(N, \boldsymbol{C}) | \det |U| = 1\} \tag{2.72}$$

If this is limited to the execution column, it is generally defined as

$$\mathrm{SU}(N, \boldsymbol{R}) \equiv \{U \in \mathrm{GU}(N, \boldsymbol{R}) | \det |U| = 1\} \tag{2.73}$$

3

Group Theory and Structural Mechanics

In the previous chapter, the definition, concept, and expression of the basic group which is the basic knowledge of group theory, were described. In this chapter, we describe a generalization of symmetry conditions in equations of motion and equilibrium. That is to say, the invariant condition in which the equation, which converted the governing equation of the discrete system structure with the group G invariance to the canonical coordinate system which adapts to the geometry characteristic of the structure, becomes commutative is described.

3.1 Group Theory and Dynamics

In this section, we describe a general theory of the block diagonalization method of equations of motion for systems with geometrical symmetry. Consider Lagrangian at time t in a N discrete system with degrees of freedom in the following:

$$L \equiv L(\boldsymbol{f}(t), \boldsymbol{u}(t), \dot{\boldsymbol{u}}(t), \ddot{\boldsymbol{u}}(t), \cdots) \tag{3.1}$$

where $\boldsymbol{f} \in \mathbf{R}^N$ and $\boldsymbol{u} \in \mathbf{R}^N$ represent the external force and displacement vector, respectively, and $\dot{\boldsymbol{u}} = d\boldsymbol{u}/dt$, $\ddot{\boldsymbol{u}} = d^2\boldsymbol{u}/dt^2$, $\cdots$. The following Lagrange's equation of motion, which is a system of differential equations, can be obtained from the minimum condition for the variant on $\boldsymbol{u}$ in Eq.(3.1).

$$\boldsymbol{F}(\boldsymbol{f}, \boldsymbol{u}, \dot{\boldsymbol{u}}, \cdots) \equiv \left(\frac{\partial L}{\partial \boldsymbol{u}} + \frac{d}{dt} \frac{\partial L}{\partial \dot{\boldsymbol{u}}} + \cdots \right)^{\mathrm{T}} = \boldsymbol{0} \tag{3.2}$$

In this book, in particular, we will consider equations of the following form

$$\boldsymbol{F} = \sum_{j=0} \Gamma_j \frac{d^j \boldsymbol{u}}{dt^j} - \boldsymbol{f} = \boldsymbol{0} \tag{3.3}$$

where, $\Gamma_0 = K \in \mathbf{R}^{N \times N}$ is the stiffness matrix, $\Gamma_1 = C \in \mathbf{R}^{N \times N}$ is the damping matrix, $\Gamma_2 = M \in \mathbf{R}^{N \times N}$ is the mass matrix, etc. Also, let $d^0/dt^0 = 1$.

In describing the symmetry of the equation of motion, we consider the group G, which consists of the source g representing geometric transformations. For example, suppose that the source g of the group G acts on the

DOI: 10.1201/9781032670386-3

N-dimensional vector $\boldsymbol{u}(t)$, and $\boldsymbol{u}(t)$ is transformed to $g(\boldsymbol{u}(t))$. The representation matrix $T(g)$ of $N \times N$, which represents the mechanism of this coordinate transformation, satisfies the following equation.

$$T(g)\boldsymbol{u}(t) = g(\boldsymbol{u}(t)), \quad {}^{\forall}g \in G \tag{3.4}$$

In this book, $\boldsymbol{f}$ and $\boldsymbol{u}$ are assumed to exist in the same space, so the representation matrix of the external force vector $\boldsymbol{f}$ is also the same.

The invariance of the Lagrangian L (group G-invariance) expresses the symmetry of this system as

$$L(T(g)\boldsymbol{f}, T(g)\boldsymbol{u}, T(g)\dot{\boldsymbol{u}}, \cdots) = L(\boldsymbol{f}, \boldsymbol{u}, \dot{\boldsymbol{u}}, \cdots), \quad {}^{\forall}g \in G \tag{3.5}$$

The coordinate transformation $T(g)$ is induced by the source g of the group G. Taking the variates with respect to $\boldsymbol{u}$ for both sides of Eq.(3.5), multiplying by $T(g)$ from the left, and using $T(g)T(g)^{\mathrm{T}} = \mathrm{I}$, we obtain the invariant conditional expression for the equation of motion $\boldsymbol{F}$,

$$T(g)\boldsymbol{F}(\boldsymbol{f}, \boldsymbol{u}, \dot{\boldsymbol{u}}, \cdots) = \boldsymbol{F}(T(g)\boldsymbol{f}, T(g)\boldsymbol{u}, T(g)\dot{\boldsymbol{u}}, \cdots), \quad {}^{\forall}g \in G \tag{3.6}$$

The conditional expression (3.6) is a generalization of the geometric symmetry condition, which states that transforming the variables $\boldsymbol{f}$ and $\boldsymbol{u}, \dot{\boldsymbol{u}}, \cdots$ by $T(g)$, respectively, is equivalent to transforming the entire expression $\boldsymbol{F}$ by $T(g)$. When written down using Eq.(3.2), Eq.(3.6) becomes

$$\sum_{j=0} T(g)\Gamma_j \frac{d^j \boldsymbol{u}}{dt^j} - T(g)\boldsymbol{f} = \sum_{j=0} \Gamma_j T(g) \frac{d^j \boldsymbol{u}}{dt^j} - T(g)\boldsymbol{f}, \quad {}^{\forall}g \in G \tag{3.7}$$

Since this expression holds for all $d^j \boldsymbol{u}/dt^j$ $(j = 0, 1, \cdots)$, we can find the symmetry condition for each matrix Γ_j as

$$T(g)\Gamma_j = \Gamma_j T(g), \quad j = 0, 1, \cdots, \quad {}^{\forall}g \in G \tag{3.8}$$

Equation (3.7) shows that the matrix $\Gamma_j(j = 0, 1, \cdots)$ can be block-diagonalized simultaneously by the same appropriate coordinate transformation matrix [27, 30].

3.2 Dynamic Problem

The consistency of various discretization analysis methods with block diagonalization theory is possible when the generalized conditionals (3.6) and (3.21) hold, as discussed in Section 3.1. This section is an extension of block diagonalization theory to D_n-invariant differential equations, complex dynamic stiffness matrices and matrices of control systems. Let $R(G)$ be the set of irreducible representations of a group G, and define the coordinate system corresponding to the irreducible representations as follows:

$$\boldsymbol{u}(t) = H\boldsymbol{w}(t) - \sum_{\mu \in R(G)} H^\mu \boldsymbol{w}^\mu(t)$$

where the coordinate transformation matrix H and the new coordinate system variables $\boldsymbol{w}$ are

$$H = [\cdots, H^\mu, \cdots],$$
$$\boldsymbol{w} = [\cdots, (\boldsymbol{w}^\mu)^{\mathrm{T}}, \cdots]^{\mathrm{T}}, \quad {}^\forall\mu \in R(G) \qquad (3.9)$$

can be expressed as a variable for each irreducible representation.

The coordinate transformation matrix H allows us to simultaneously calculate

$$\widetilde{\Gamma}_j = H^{\mathrm{T}}\Gamma_j H = \mathrm{diag}[\cdots, \widetilde{\Gamma}_j^\mu, \cdots] \qquad (3.10)$$

and block diagonalize the matrix $\Gamma_j (j = 0, 1, \cdots)$ of the differential equation system. Where $\mathrm{diag}[\cdots]$ denotes the block diagonal matrix and is

$$\widetilde{\Gamma}_j^\mu = (H^\mu)^{\mathrm{T}}\Gamma_j H^\mu, \quad {}^\forall\mu \in R(G). \qquad (3.11)$$

The block $\widetilde{\Gamma}_j^\mu$ corresponding to the d-ordered irreducible representation ($d \geq 2$) has a detailed block diagonal structure consisting of an additional d identical blocks.

Transforming the left-hand side of the equation of motion (3.2) with the coordinate transformation matrix H and using equations (3.9) and (3.10) yields

$$\begin{aligned} H^{\mathrm{T}}\boldsymbol{F} \quad &= \sum_{j=0} H^{\mathrm{T}}\Gamma_j \frac{d^j H\boldsymbol{w}}{dt^j} - H^{\mathrm{T}}\boldsymbol{f} \\ &= \sum_{j=0} (H^{\mathrm{T}}\Gamma_j H)\frac{d^j \boldsymbol{w}}{dt^j} - H^{\mathrm{T}}\boldsymbol{f} \\ &= \sum_{j=0} \widetilde{\Gamma}_j \frac{d^j \boldsymbol{w}}{dt^j} - H^{\mathrm{T}}\boldsymbol{f} \end{aligned} \qquad (3.12)$$

but, since $\widetilde{\Gamma}_j$ is block-diagonal, the equation of motion (3.2) can be decomposed into

$$\sum_{j=0} \widetilde{\Gamma}_j^\mu \frac{d^j \boldsymbol{w}^\mu}{dt^j} - (H^\mu)^{\mathrm{T}}\boldsymbol{f} = \boldsymbol{0}, \quad {}^\forall\mu \in R(G) \qquad (3.13)$$

and independent differential equations in the form corresponding to each irreducible expression (the equation corresponding to the d-order irreducible expression ($d \geq 2$) can be further decomposed into d pieces). Substituting the initial conditions for each irreducible representation into this differential equation, the solution $\boldsymbol{u}$ of the original equation is obtained from Eq.(3.9) as a superposition of the solutions $\boldsymbol{w}^\mu$.

In solving the equation of motion (3.2), assuming that the general solution is the harmonic system $\boldsymbol{u}(t) = \bar{\boldsymbol{u}}\exp(\lambda t)$, Eq.(3.2) becomes

$$\boldsymbol{F} = S(\lambda)\bar{\boldsymbol{u}} - \boldsymbol{f} = \boldsymbol{0} \qquad (3.14)$$

Here, $S(\lambda) = \sum_{j=0} \lambda^j \Gamma_j$ is a complex dynamic stiffness matrix, $\bar{\boldsymbol{u}}$ is an initial value vector, and λ is a complex number.

Using the coordinate transformation matrix H (refer to Eq.(3.10)), the complex dynamic stiffness matrix $S(\lambda)$ can be block-diagonalized in the following:

$$\begin{aligned}
\widetilde{S}(\lambda) &= H^{\mathrm{T}} S(\lambda) H = \sum_{j=0} \lambda^j H^{\mathrm{T}} \Gamma_j H \\
&= \mathrm{diag}[\cdots, \widetilde{S}^\mu, \cdots] \qquad (3.15)
\end{aligned}$$

Here,

$$\widetilde{S}^\mu(\lambda) = (H^\mu)^{\mathrm{T}} S(\lambda) H^\mu = \sum_{j=0} \lambda^j \widetilde{\Gamma}_j^\mu \qquad (3.16)$$

Since H is an orthogonal matrix, the eigenvalues of the transformed complex dynamic stiffness matrix $\widetilde{S}(\lambda)$ are identical to the eigenvalues of $S(\lambda)$. Furthermore, since $\widetilde{S}(\lambda)$ is a block diagonal, it is to our advantage to replace the eigenvalue analysis of $S(\lambda)$ with the eigenvalue analysis of each block $\widetilde{S}^\mu(\lambda)$ in order to improve the efficiency and stability of the numerical analysis. This is highly beneficial in terms of the efficiency and stability of the numerical analysis.

3.3 Group Theory and Statics

In this section, we discuss how to describe the symmetry of a system of static equilibrium equations with geometric symmetry. Total potential energy of a discrete system is $\mathcal{U}(\boldsymbol{f}, \boldsymbol{u})$. The total potential energy is used here, but the following discussion holds for systems with no potential. $\boldsymbol{f}$ and $\boldsymbol{u}$ denote the load pattern vector and displacement vector, respectively. The equilibrium equation (N-dimension) of this system is

$$\boldsymbol{F}(\boldsymbol{f}, \boldsymbol{u}) \equiv \left(\frac{\partial \mathcal{U}}{\partial \boldsymbol{u}}\right)^{\mathrm{T}} = \boldsymbol{0} \qquad (3.17)$$

To describe the symmetry of the equilibrium equation, we assume a group G consisting of transformation maps g (representing mirroring, rotation, etc.).

For example, suppose that the source g of the group G acts on the N-dimensional vector $\boldsymbol{u}$ (or $\boldsymbol{F}$), which transforms $\boldsymbol{u}$ into $g(\boldsymbol{u})$ (or $\boldsymbol{F}$ into $g(\boldsymbol{F})$). The representation matrix $T(g)$ of $N \times N$, which represents the mechanism of this coordinate transformation, is the one that satisfies

$$T(g)\boldsymbol{u} = g(\boldsymbol{u}), \quad T(g)\boldsymbol{F} = g(\boldsymbol{F}), \quad g \in G. \tag{3.18}$$

Similarly, the representation matrix of the load vector $\boldsymbol{f}$ [1] is defined by

$$\widetilde{T}(g)\boldsymbol{f} = g(\boldsymbol{f}), \quad g \in G. \tag{3.19}$$

The representation matrix is a representation of the coordinate transformation induced by the source g in terms of the corresponding vector space, and in this book, $T(g)$ and $\widetilde{T}(g)$ are assumed to be orthogonal matrices.

The symmetry of this system is expressed by the invariance of the total potential energy $\mathcal{U}$ to the coordinate transformations induced by the source g. We say that the total potential energy $\mathcal{U}$ is invariant to the group G if the following equation holds for all sources g of the group G;

$$\mathcal{U}(\widetilde{T}(g)\boldsymbol{f}, T(g)\boldsymbol{u}) = \mathcal{U}(\boldsymbol{f}, \boldsymbol{u}), \quad g \in G \tag{3.20}$$

In this case,

$$T(g)\boldsymbol{F}(\boldsymbol{f}, \boldsymbol{u}) = \boldsymbol{F}(\widetilde{T}(g)\boldsymbol{f}, T(g)\boldsymbol{u}), \quad g \in G \tag{3.21}$$

stands for the equilibrium equation. Eq.(3.21) is a general symmetry condition that can be considered for non-potential systems and is called the identity for the group G of equilibrium equations $\boldsymbol{F}$. The conditional expression (3.21) is a generalization of the geometric symmetry condition, which states that the transformation of the variables $\boldsymbol{f}$ and $\boldsymbol{u}$ by $\widetilde{T}(g)$ and $T(g)$, respectively, is identical to the transformation of the entire expression $\boldsymbol{F}$ by $T(g)$.

The stiffness matrix K of the linear stiffness equation

$$\boldsymbol{F} \equiv K\boldsymbol{u} - \boldsymbol{f} = 0 \tag{3.22}$$

in this book satisfies the symmetry condition

$$T(g)K = KT(g), \quad g \in G \tag{3.23}$$

according to Eq.(3.21), and can be block-diagonalized by a suitable coordinate transformation. [2]

It is known that an equilibrium equation with the same degeneration (3.21) for a group can be decomposed into an equation corresponding to the irreducible representation of the group by appropriate coordinate transformations. Which coordinate transformation matrices are taken and how they are decomposed are individual theories for each group.

---

[1]Since the load vector $\boldsymbol{f}$ may have a different dimension than $\boldsymbol{u}$ or $\boldsymbol{F}$, the representation matrix of $\boldsymbol{f}$ is generally different from the representation matrix of $\boldsymbol{u}$ or $\boldsymbol{F}$.

[2]Since the symmetry condition in Eq.(3.23) holds for the tangent stiffness matrix of a nonlinear problem, the argument of this book holds [27, 30].

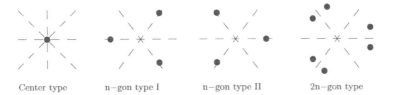

Center type   n−gon type I   n−gon type II   2n−gon type

FIGURE 3.1
4 kinds of orbits

3.4 Orbital Concept

3.4.1 Orbital Concept and Matrix Coordinate Transforms

The set of nodes of a discretized structural system that is D_n-invariant is also D_n-invariant and can be decomposed into a minimal unit called (D_n-invariant) orbit. This trajectory represents the set of points $r^k(\boldsymbol{x})$ and $sr^k(\boldsymbol{x})$ transformed from a node $\boldsymbol{x}$ by elements r^k and sr^k ($k = 0, 1, \cdots, n-1$) of D_n and can be defined as

$$\{r^k(\boldsymbol{x}),\ sr^k(\boldsymbol{x}) \in \mathbf{R}^2 \mid k = 0, 1, \cdots, n-1\} \tag{3.24}$$

Due to the coordinate transformation caused by D_n, the orbit as a whole remains invariant even if individual nodes are moved. The orbits of the nodes of D_n-invariant discrete systems can be classified into four types,

$$\text{type of orbits} \begin{cases} \text{Center type} \\ n - \text{gon type} \\ n - \text{gon type} \\ 2n - \text{gon type} \end{cases}$$

as in **Fig. 3.1**. When the column vector of the coordinate transformation matrix H is determined for each orbit, the matrix H can be taken sparsely, which increases the efficiency of calculation and has the advantage that the calculation of H can be handled systematically.

Substituting the formula for calculating H_v for each orbit proposed by [27, 28] into the right-hand side of Eq.(5.50), we can obtain the coordinate transformation matrix H_θ of the rotational displacement.

The partial block matrix of H for a structural system of N_O orbitals is very sparse in the form of

$$H^\mu = \text{diag}[H_1^\mu, \cdots, H_{No}^\mu]$$

$$= \begin{pmatrix} H_1^\mu & O & O & O & O \\ O & \cdot & O & O & O \\ O & O & \cdot & O & O \\ O & O & O & \cdot & O \\ O & O & O & O & H_{No}^\mu \end{pmatrix}, \quad \mu \in R(G) \qquad (3.25)$$

which only has a component for each orbital. It is worth noting that it is clear from the Eq.(5.45) that the representation matrices of the translational displacement v and rotational displacement θ in Section 5.3 are independent, and the representation matrices of the X, Y and Z direction components of these two representations are also independent. Corresponding to this independence of the representation matrices, each block H_q^μ ($q = 1, \cdots, N_O$) of the coordinate transformation matrix has an additional sparse detail block structure called

$$H_q^\mu = \text{diag}[H_{q,v_{XY}}^\mu, H_{q,v_Z}^\mu, H_{q,\theta_{XY}}^\mu, H_{q,\theta_Z}^\mu]$$

$$= \begin{pmatrix} H_{q,v_{XY}}^\mu & O & O & O \\ O & H_{q,v_Z}^\mu & O & O \\ O & O & H_{q,\theta_{XY}}^\mu & O \\ O & O & O & H_{q,\theta_Z}^\mu \end{pmatrix}, \quad \mu \in R(G) \qquad (3.26)$$

In calculating the block matrix of the stiffness matrix $\widetilde{K}^\mu$ by the formula (5.19), it is highly advantageous to use the sparsity of the H matrix, as in formulas (5.54) and (5.55). For an efficient method of computing the block matrix $\widetilde{K}^\mu$ of the stiffness matrix, see Appendix or [31].

4

Basics of the Finite Element Method

Generally, displacements and stresses in a structure are changeable into complex forms and become complex functions involving the coordinates (x, y, and z). However, if only focused on a small area in a structure, the displacement and stress within the area can be approximated by a simple function.

For example, **Fig. 4.1**(a) is a phenomenon after cylindrical buckling due to Origami, which became famous in the paper [41]. **Fig. 4.1**(b) can be obtained by "finite element method." Alternatively, as shown in **Fig. 4.1**(c), approximate analysis can be performed as a truss model that extracts only the uneven part of the buckling deformation. Ultimately, in order to have a computer calculate the mechanical analysis of a structure, a structural model for approximate calculation consisting of nodes and various elements must be considered.

4.1 Finite Element Method as a Structural Analysis for a Discrete Model

There are various ways to express the displacement in the element from the displacement at the nodal point. In the finite element method, it is generally

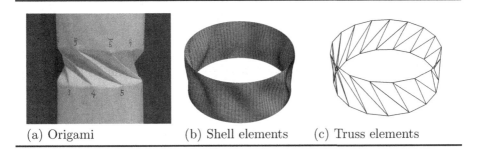

| (a) Origami | (b) Shell elements | (c) Truss elements |

FIGURE 4.1
Finite element division of symmetric structure

DOI: 10.1201/9781032670386-4   43

expressed by a simple interpolation function. From this, if the displacement at the nodal point is clarified, then the displacement of any point in the structure can be obtained by this interpolation function. Moreover, if the displacement becomes clear by using this method, the stress can also be obtained.

About Node Displacements

The of each element is expressed as the nodal displacements as an unknown quantity, and the is applied. Alternatively, the virtual work principle is applied to the virtual work equation of virtual strain energy and load due to the virtual nodal displacements to derive the equilibrium equation of the elements. This equilibrium equation is a system of algebraic equations for nodal displacements. Assemble this for the entire structure to create an equilibrium equation for the entire structure, and solve this equation to obtain the node displacements.

Characteristics of the Finite Element Method

- Applicable to any type of structure.

- Applicable not only to linear analysis but also to non-linear analysis such as buckling analysis and elasto-plastic analysis.

- Not limited to structural mechanics problems, but also applicable to problems such as water flow, water penetration, and heat transfer.

- Much general-purpose finite element analysis software has been developed.

- The result of the Finite Element Method is an approximate solution depending on the element division.

4.2   Fundamentals of Continuum Mechanics

The strain-displacement relational expression, stress-strain relational expression, and total potential energy of solid members, plate and shell members, and frame members are shown. Worthly of note, the material is a homogeneous isotropic linear elastic body.

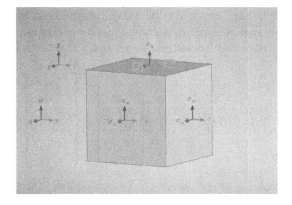

FIGURE 4.2
Displacement and stress of solid members

4.2.1 Solid Members

When the x, y, z axial displacement is expressed by u, v, w, the strain is expressed from the displacement as follows:

$$
\left\{
\begin{array}{c}
\varepsilon_{xx} \\
\varepsilon_{yy} \\
\varepsilon_{xy} \\
2\varepsilon_{xy} \\
2\varepsilon_{yz} \\
2\varepsilon_{zx}
\end{array}
\right\}
=
\left(
\begin{array}{ccc}
\frac{\partial}{\partial x} & 0 & 0 \\
0 & \frac{\partial}{\partial y} & 0 \\
0 & 0 & \frac{\partial}{\partial z} \\
\frac{\partial}{\partial y} & \frac{\partial}{\partial x} & 0 \\
0 & \frac{\partial}{\partial z} & \frac{\partial}{\partial y} \\
\frac{\partial}{\partial z} & 0 & \frac{\partial}{\partial x}
\end{array}
\right)
\left\{
\begin{array}{c}
u \\
v \\
w
\end{array}
\right\}
$$

$$
\varepsilon = Lu \tag{4.1}
$$

In addition, the relationship between strain and displacement as in the above equation, and the equation are simply expressed as $\varepsilon = Lu$ by matrix form. In an isotropic elastic body, the stress is given by the following equation from the strain;

$$
\left\{
\begin{array}{c}
\sigma_{xx} \\
\sigma_{yy} \\
\sigma_{zz} \\
\sigma_{xy} \\
\sigma_{yz} \\
\sigma_{zx}
\end{array}
\right\}
=
\left(
\begin{array}{ccc|ccc}
\lambda + 2G & \lambda & \lambda & & & \\
\lambda + 2G & \lambda & & & O & \\
\lambda & \lambda + 2G & & & & \\
\hline
& & & G & & \\
& O & & & G & \\
& & & & & G
\end{array}
\right)
\left\{
\begin{array}{c}
\varepsilon_{xx} \\
\varepsilon_{yy} \\
\varepsilon_{xy} \\
2\varepsilon_{xy} \\
2\varepsilon_{yz} \\
2\varepsilon_{zx}
\end{array}
\right\} \tag{4.2}
$$

where, λ is the rigid frame constant, G is the shear modulus, and the following equation can be obtained from Young's modulus E and Poisson's ratio ν;

$$
\lambda = \frac{\nu E}{(1 + \nu)(1 - 2\nu)}, \quad G = \frac{E}{2(1 + \nu)} \tag{4.3}
$$

In addition, the relational expression between stress and strain, as in the above equation is simply expressed as $\boldsymbol{\sigma} = D\boldsymbol{\varepsilon}$ by matrix notation.

The following equation from stress and strain gives the strain energy $\mathcal{U}$.

$$\mathcal{U} = \frac{1}{2}\int_V \boldsymbol{\sigma}^{\mathrm{T}} \boldsymbol{\varepsilon}\, dV = \frac{1}{2}\int_V \begin{Bmatrix} \sigma_{xx} \\ \sigma_{yy} \\ \sigma_{zz} \\ \sigma_{xy} \\ \sigma_{yz} \\ \sigma_{zx} \end{Bmatrix}^{\mathrm{T}} \begin{Bmatrix} \varepsilon_{xx} \\ \varepsilon_{yy} \\ \varepsilon_{xy} \\ 2\varepsilon_{xy} \\ 2\varepsilon_{yz} \\ 2\varepsilon_{zx} \end{Bmatrix} dV \qquad (4.4)$$

where, V is the volume of the member.

The load acting on the inside of the member is $\boldsymbol{p} = \{p_x, p_y, p_z\}^{\mathrm{T}}$, and the load acting on the surface of the member is $\boldsymbol{P} = \{P_x, P_y, P_z\}^{\mathrm{T}}$. Then, the work $\mathcal{W}$ by load is given by the following equation.

$$\mathcal{W} = \int_V \boldsymbol{p}^{\mathrm{T}}\boldsymbol{u}\, dV + \int_{S_\sigma} \overline{\boldsymbol{P}}^{\mathrm{T}}\boldsymbol{u}\, dS \qquad (4.5)$$

where, S_σ represents the area on the surface where the load is applied.

From there, the total potential energy Π is defined by the following equation.

$$\Pi = \mathcal{U} - \mathcal{W} = \Pi(\boldsymbol{u}) \qquad (4.6)$$

4.2.2   Plate and Shell elements

For thin plate-shaped members such as decks, walls, and roofs, the strain and stress in the thickness direction are smaller than the strain and stress in the plate surface, and may be ignored in engineering. Such a member should be treated as a plate and shell element using some assumptions made by taking advantage of the characteristics of its structural shape rather than a solid element. The following assumptions can be made for plate and/or shell elements.

1. No deformation in the thickness direction. ($\varepsilon_{zz} = 0$)

2. No stress in the thickness direction. ($\sigma_{zz} = 0$)

3. The cross-section perpendicular to the plane of the plate keeps an even plane after deformation.($2\varepsilon_{yz}$ and $2\varepsilon_{zx}$ are constant $z-$ direction)

Theory based on these assumptions is called **Mindlin's plate theory**. Instead of Mindlin's plate theory assumption (3), the following assumption is called **Kirchhoff-Love's plate theory**.

(3) Initially, the cross-section that was perpendicular to the plane of $z = 0$ (center plane of plate thickness) is perpendicular to the center plane of the plate after deformation. ($2\varepsilon_{yz} = 0, 2\varepsilon_{zx} = 0$) If the plate thickness is very thin,

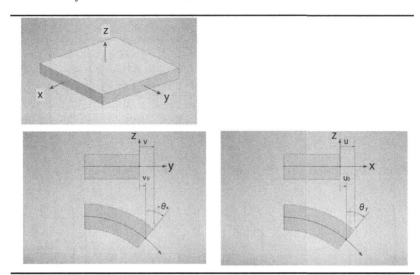

FIGURE 4.3
Displacements of a plate element

this is also a valid assumption. However, when the plate thickness becomes relatively thick, the solution based on this assumption does not represent the actual behavior, so Kirchhoff-Love's plate theory is **thin plate theory**, and Mindlin's plate theory is **thin plate theory**. The coordinate system is set so that the z- axis, x, and y axes are in the plane of the plate in the thickness direction.

4.2.3 Mindlin's Plate Theory

From assumptions (1) and (3), the displacements u, v, and w in the member are the displacements u_0, v_0, and w_0 in the plane of $z = 0$. And, the rotation of the cross-section around the x and y axes can be represented as θ_x and θ_y and is expressed as follows:

$$u = u_0 + z\theta_y, \ v = v_0 - z\theta_x, \ w = w_0 \tag{4.7}$$

Substituting this into the strain Eq.(4.1) for solid elements, the relational equation between strain and displacement of plate and shell elements is as follows:

$$
\left\{
\begin{array}{c}
\varepsilon_{xx} \\
\varepsilon_{yy} \\
2\varepsilon_{xy} \\
2\varepsilon_{yz} \\
2\varepsilon_{zx}
\end{array}
\right\}
=
\left(
\begin{array}{ccc|cc}
\frac{\partial}{\partial x} & 0 & 0 & 0 & z\frac{\partial}{\partial x} \\
0 & \frac{\partial}{\partial y} & 0 & z\frac{\partial}{\partial y} & 0 \\
\frac{\partial}{\partial y} & \frac{\partial}{\partial x} & 0 & z\frac{\partial}{\partial x} & z\frac{\partial}{\partial y} \\
0 & 0 & \frac{\partial}{\partial y} & -1 & 0 \\
0 & 0 & \frac{\partial}{\partial x} & 0 & 1
\end{array}
\right)
\left\{
\begin{array}{c}
u_0 \\
v_0 \\
w_0 \\
\theta_x \\
\theta_y
\end{array}
\right\}
$$

$$\varepsilon \;=\; L\hat{u} \tag{4.8}$$

From assumption (2), the relational expression between stress and strain is as follows:

$$
\left\{
\begin{array}{c}
\sigma_{xx} \\
\sigma_{yy} \\
\sigma_{xy} \\
\sigma_{yz} \\
\sigma_{zx}
\end{array}
\right\}
=
\frac{E}{1-\nu^2}
\left(
\begin{array}{ccc|cc}
1 & \nu & 0 & 0 & 0 \\
\nu & 1 & 0 & 0 & 0 \\
0 & 0 & \frac{1-\nu}{2} & 0 & 0 \\
0 & 0 & 0 & \kappa\frac{1-\nu}{2} & 0 \\
0 & 0 & 0 & 0 & \kappa\frac{1-\nu}{2}
\end{array}
\right)
\left\{
\begin{array}{c}
\varepsilon_{xx} \\
\varepsilon_{yy} \\
2\varepsilon_{xy} \\
2\varepsilon_{yz} \\
2\varepsilon_{zx}
\end{array}
\right\}
$$

$$\sigma \;=\; D\varepsilon \tag{4.9}$$

Here, κ is a coefficient for correcting the contribution of shear deformation to the strain energy. It is called the **shear correction factor** and $\kappa = 5/6$ is used.

The strain energy $\mathcal{U}$ is expressed as follows:

$$
\mathcal{U} = \frac{1}{2} \int_V
\left\{
\begin{array}{c}
\sigma_{xx} \\
\sigma_{yy} \\
\sigma_{xy} \\
\sigma_{yz} \\
\sigma_{zx}
\end{array}
\right\}^{\mathrm{T}}
\left\{
\begin{array}{c}
\varepsilon_{xx} \\
\varepsilon_{yy} \\
2\varepsilon_{xy} \\
2\varepsilon_{yz} \\
2\varepsilon_{zx}
\end{array}
\right\}
dV
\tag{4.10}
$$

The load acting on the inside of the member is $p = \{p_x, p_y, p_z, m_x, m_y\}^{\mathrm{T}}$, and the load acting on the surface is $P = \{P_x, P_y, P_z\}^{\mathrm{T}}$. The work $\mathcal{W}$ by load is given by the following equation.

$$\mathcal{W} = \int_A p^{\mathrm{T}}\hat{u}\, dA + \int_{S_\sigma} \overline{P}^{\mathrm{T}}\hat{u}\, dS \tag{4.11}$$

where, A represents the area of the central surface of the plate and shell members, and these represent the area on the periphery of the member on which the load is applied. From these, the total potential energy Π is defined by the following equation;

$$\Pi = \mathcal{U} - \mathcal{W} = \Pi(\hat{u}) \tag{4.12}$$

In the plate and shell theory, instead of stress, the following combined stress is used after the integral of the stress in the z-axis direction.

$$N_{xx} = \int_{-h/2}^{h/2} \sigma_{xx} dz \ , \ N_{yy} = \int_{-h/2}^{h/2} \sigma_{yy} dz \ , \ N_{xy} = \int_{-h/2}^{h/2} \sigma_{xy} dz \ ,$$

$$Q_{yz} = \int_{-h/2}^{h/2} \sigma_{yz} dz \ , \ Q_{zx} = \int_{-h/2}^{h/2} \sigma_{zx} dz \ ,$$

$$M_{xx} = \int_{-h/2}^{h/2} \sigma_{xx} z dz \ , \ M_{yy} = \int_{-h/2}^{h/2} \sigma_{yy} z dz \ , \ M_{xy} = \int_{-h/2}^{h/2} \sigma_{xy} z dz$$

Note that, h is the plate thickness. The relational expression between the combined stress and the displacement is as follows:

$$
\begin{aligned}
N_{xx} &= \frac{Eh}{1-\nu^2}\left(\frac{\partial u_0}{\partial x} + \nu \frac{\partial v_0}{\partial y}\right), & M_{xx} &= \frac{Eh^3}{12(1-\nu^2)}\left(\frac{\partial \theta_y}{\partial x} - \nu \frac{\partial \theta_x}{\partial y}\right) \\
N_{yy} &= \frac{Eh}{1-\nu^2}\left(\nu \frac{\partial u_0}{\partial x} + \frac{\partial v_0}{\partial y}\right), & M_{yy} &= \frac{Eh^3}{12(1-\nu^2)}\left(\nu \frac{\partial \theta_y}{\partial x} - \frac{\partial \theta_x}{\partial y}\right) \\
N_{xy} &= Gh\left(\frac{\partial u_0}{\partial y} + \frac{\partial v_0}{\partial x}\right), & M_{xy} &= \frac{Gh^3}{12}\left(\frac{\partial \theta_y}{\partial y} - \nu \frac{\partial \theta_x}{\partial x}\right) \\
Q_{yz} &= \kappa Gh\left(-\theta_x + \frac{\partial w_0}{\partial y}\right), & Q_{zx} &= \kappa Gh\left(\frac{\partial w_0}{\partial x} + \theta_y\right)
\end{aligned}
$$

The strain energy is expressed as follows:

$$
\begin{aligned}
\mathcal{U} = \ & \frac{1}{2}\int_A \left\{ N_{xx}\frac{\partial u_0}{\partial x} + N_{yy}\frac{\partial v_0}{\partial y} + N_{xy}\left(\frac{\partial u_0}{\partial y} + \frac{\partial v_0}{\partial x}\right) \right. \\
& + Q_{yz}\left(\frac{\partial w_0}{\partial y} - \theta_x\right) + Q_{zx}\left(\frac{\partial w_0}{\partial x} + \theta_y\right) \\
& \left. + M_{xx}\frac{\partial \theta_y}{\partial x} - M_{yy}\frac{\partial \theta_x}{\partial y} + M_{xy}\left(\frac{\partial \theta_y}{\partial y} - \frac{\partial \theta_x}{\partial x}\right) \right\} dA
\end{aligned}
$$

4.2.4 Kirchhoff-Love's Plate Theory

Instead for assumption (3), using assumptions (3)', the following relationship is established by deflection and rotation angle.

From this, the relationship between strain and displacement, and the equation is as follows:

$$
\left\{\begin{array}{c} \varepsilon_{xx} \\ \varepsilon_{yy} \\ 2\varepsilon_{xy} \end{array}\right\} = \left(\begin{array}{ccc} \frac{\partial}{\partial x} & 0 & -z\frac{\partial^2}{\partial x^2} \\ 0 & \frac{\partial}{\partial y} & -z\frac{\partial^2}{\partial y^2} \\ \frac{\partial}{\partial y} & \frac{\partial}{\partial x} & -2z\frac{\partial^2}{\partial x \partial y} \end{array}\right)\left\{\begin{array}{c} u_0 \\ v_0 \\ w_0 \end{array}\right\}
$$

$$\varepsilon = \boldsymbol{L}\boldsymbol{u} \tag{4.13}$$

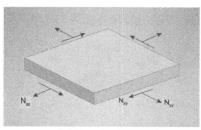

(a) In-plane force

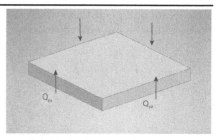

(b) Out-of-plane shear force

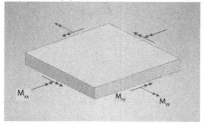

(c) Out-of-plane bending moment

FIGURE 4.4
Combined stress for a plate

The relational expression between stress and strain is as follows:

$$\left\{ \begin{array}{c} \sigma_{xx} \\ \sigma_{yy} \\ \sigma_{xy} \end{array} \right\} = \frac{E}{1-\nu^2} \left(\begin{array}{ccc} 1 & \nu & 1 \\ \nu & 1 & 0 \\ 0 & 0 & \frac{1-\nu}{2} \end{array} \right) \left\{ \begin{array}{c} \varepsilon_{xx} \\ \varepsilon_{yy} \\ 2\varepsilon_{xy} \end{array} \right\}$$

$$\boldsymbol{\sigma} = D\boldsymbol{\varepsilon} \qquad (4.14)$$

The strain energy $\mathcal{U}$ is expressed as follows:

$$\mathcal{U} = \frac{1}{2} \int_V \left\{ \begin{array}{c} \sigma_{xx} \\ \sigma_{yy} \\ \sigma_{xy} \\ \sigma_{yz} \\ \sigma_{zx} \end{array} \right\}^{\mathrm{T}} \left\{ \begin{array}{c} \varepsilon_{xx} \\ \varepsilon_{yy} \\ 2\varepsilon_{xy} \\ 2\varepsilon_{yz} \\ 2\varepsilon_{zx} \end{array} \right\} dV \qquad (4.15)$$

The load acting on the inside of the member is $\boldsymbol{p} = \{p_x, p_y, p_z, m_x, m_y\}^{\mathrm{T}}$, and the load acting on the surface is $\boldsymbol{P} = \{\overline{P}_x, \overline{P}_y, \overline{P}_z, \overline{M}_x, \overline{M}_y\}^{\mathrm{T}}$, the work $\mathcal{W}$ by load is given by the following equation.

$$\mathcal{W} = \int_V \boldsymbol{p}^{\mathrm{T}} \boldsymbol{u} \, dA + \int_{S_\sigma} \overline{\boldsymbol{P}}^{\mathrm{T}} \boldsymbol{u} \, dS \qquad (4.16)$$

Here, A represents the area of the central surface of the plate and shell member, and S_σ represents the area around the member on which the load is applied.

From these, the total potential energy Π is defined by the following equation;

$$\Pi = \mathcal{U} - \mathcal{W} = \Pi(\boldsymbol{u}) \tag{4.17}$$

4.3 Rigid Frame Members

For long and narrow members such as rods, the stress and strain in the cross-section are smaller than the stress and strain in the longitudinal direction, and may be neglected in engineering. It is better to treat such a member as a rigid frame member with certain assumptions by taking advantage of its structural shape characteristics rather than using a solid member. For rigid frame members, the assumptions are as follows:

- (1) Cross-section does not deform. $\varepsilon_{yy} = 0, \varepsilon_{zz} = 0, \varepsilon_{yz} = 0$

- (2) Stress in cross-section is zero. $\varepsilon_{yy} = 0, \sigma_{zz} = 0, \sigma_{xy} = 0$

- (3) Initially, the cross-section forming the normal plane of the x-axis remains flat even after deformation. The theory based on these assumptions is called **Timoshenko's beam theory**. Timoshenko's beam theory with the following assumptions is called **Bernoulli-Euler's beam theory**.

- (3)' Initially, the cross-section forming the normal plane of the x-axis remains flat after deformation and matches the normal plane of the z-axis after deformation. For very long and narrow beams, this is a very reasonable assumption. However, for relatively short beams, the solution under this assumption does not represent the actual behavior. The coordinate system, x-axis in the longitudinal direction, y-axis and z-axis is set to be in cross-section.

4.3.1 Timoshenko Beam Theory

From assumptions (1) and (3), the displacements u, v, and w in the element are the displacements u_0, v_0, and w_0 on the x-axis. From the x-axis, y-axis, and z-axis rotation angles that represent the rotation of the cross-section, it is expressed as follows from $\theta_x, \theta_y, \theta_z$;

$$u = u_0 - y\theta_z + z\theta_y \ , \quad v = v_0 - z\theta_x \ , \quad w = w_0 + y\theta_x \tag{4.18}$$

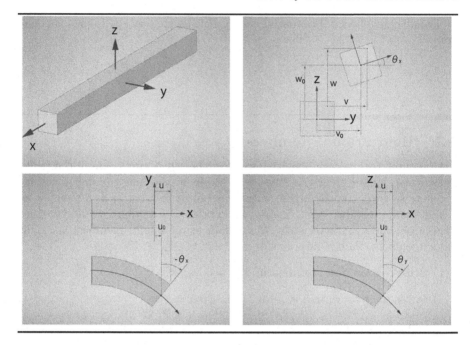

FIGURE 4.5
Rigid frame members

$$
\left\{
\begin{array}{c}
\varepsilon_{xx} \\
2\varepsilon_{xy} \\
2\varepsilon_{zx}
\end{array}
\right\}
=
\left(
\begin{array}{ccc|ccc}
\frac{d}{dx} & 0 & 0 & 0 & z\frac{d}{dx} & -y\frac{d}{dx} \\
0 & \frac{d}{dx} & 0 & -z\frac{d}{dx} & 0 & -1 \\
0 & 0 & \frac{d}{dx} & y\frac{d}{dx} & 1 & 0
\end{array}
\right)
\left\{
\begin{array}{c}
u_0 \\
v_0 \\
w_0 \\
\theta_x \\
\theta_y \\
\theta_z
\end{array}
\right\}
$$

$$\varepsilon = L\hat{u} \tag{4.19}$$

Substituting this into the strain equation for solid members (4.1), the relational equation between strain and displacement of plate and shell members is as follows: From assumption (2), the relational expression between stress and strain is

$$
\left\{
\begin{array}{c}
\sigma_{xx} \\
\sigma_{yy} \\
\sigma_{zx}
\end{array}
\right\}
=
\frac{E}{1-\nu^2}
\left(
\begin{array}{ccc}
E & 0 & 0 \\
0 & G & 0 \\
0 & 0 & G
\end{array}
\right)
\left\{
\begin{array}{c}
\varepsilon_{xx} \\
2\varepsilon_{xy} \\
2\varepsilon_{zx}
\end{array}
\right\}
$$

$$\sigma = D\varepsilon \tag{4.20}$$

The strain energy $\mathcal{U}$ is expressed as follows:

$$\mathcal{U} = \frac{1}{2} \int_V \left\{ \begin{array}{c} \sigma_{xx} \\ \sigma_{xy} \\ \sigma_{zx} \end{array} \right\}^{\mathrm{T}} \left\{ \begin{array}{c} \varepsilon_{xx} \\ 2\varepsilon_{xy} \\ 2\varepsilon_{zx} \end{array} \right\} dV \tag{4.21}$$

The load acting on the inside of the member is $p = \{p_x, p_y, p_z, m_x, m_y, m_z\}^{\mathrm{T}}$, and the load acting on the surface is $\boldsymbol{P} = \{\overline{P}_x, \overline{P}_y, \overline{P}_z, \overline{M}_x, \overline{M}_y, \overline{M}_z\}^{\mathrm{T}}$. The work $\mathcal{W}$ by load is given by the following equation.

$$\mathcal{W} = \int_0^\ell \boldsymbol{p}^{\mathrm{T}} \hat{\boldsymbol{u}} \, dx + \left[\overline{n}_x \overline{\boldsymbol{P}}^{\mathrm{T}} \hat{\boldsymbol{u}} \right]_0^\ell \tag{4.22}$$

Here, ℓ is the member length of the rigid frame member, and $\overline{n}_x$ is the direction cosine between the outward normal of the cross-section of the member end and the x-axis, $x = 0$ for $\overline{n}_x = -1$, $x = \ell$ for $\overline{n}_x = 1$.

From these, the total potential energy Π is defined by the following equation;

$$\Pi = \mathcal{U} - \mathcal{W} = \Pi(\hat{\boldsymbol{u}}) \tag{4.23}$$

In beam theory, stress is integrated within the cross-section instead of stress. The following **section force** is used;

$$N = \int_A \sigma_{xx} \, dA \,, \quad \sigma_y = \int_A \sigma_{xy} \, dA \,, \quad \sigma_z = \int_A \sigma_{zx} \, dA \,,$$

$$T = \int_A (-\sigma_{xy} z + \sigma_{zx} y) \, dA \,, \quad M_y = \int_A \sigma_{xx} z \, dA \,, \quad M_z = -\int_A \sigma_{xx} z \, dA$$

When x passes through the centroid of the cross-section and the y-axis and the z-axis point in the direction of the main axis of the cross-section, the relational expression between the cross-section force and the displacement is as follows:

$$N = EA \frac{du_0}{dx}, \quad Q_y = \kappa_y GA \left(-\theta_z + \frac{dv_0}{dx} \right), \quad Q_z = \kappa_z GA \left(\theta_y + \frac{dw_0}{dx} \right),$$

$$T = GJ \frac{d\theta_x}{dx}, \quad M_y = EI_y \frac{d\theta_y}{dx}, \quad M_z = EI_z \frac{d\theta_z}{dx} \tag{4.24}$$

Here,

$$A = \int_A dA \,, \quad I_y = \int_A z^2 \, dA \,, \quad I_z = \int_A y^2 \, dA \,, \quad J = \int_A (y^2 + z^2) \, dA$$

The strain energy is expressed as follows:

$$\mathcal{U} = \frac{1}{2} \int_0^\ell \left\{ N \frac{du_0}{dx} + \kappa_y Q_y \left(-\theta_z \frac{dv_0}{dx} \right) + \kappa_z Q_z \left(\theta_y + \frac{dw_0}{dx} \right) \right.$$

$$\left. + T \frac{d\theta_x}{dx} + M_y \frac{d\theta_y}{dx} + M_z \frac{d\theta_z}{dx} \right\} dx$$

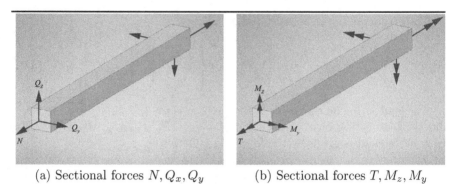

(a) Sectional forces N, Q_x, Q_y (b) Sectional forces T, M_z, M_y

FIGURE 4.6
Cross-sectional forces for a frame

Here, κ_y, κ_z are coefficients for correcting the contribution of shear deformation to strain energy. This is called the **shear correction factor**. The values of these coefficients vary depending on the shape of the cross-section, and the value of $\kappa_y = \kappa_z = 5/6$ is used for the rectangular cross-section.

4.3.2   Bernoulli-Euler Beam Theory

Using assumption (3)' instead of assumption (3), the following relationship arises between the deflection and the angle of rotation.

$$\theta_y = -\frac{dw_0}{dx}, \quad \theta_z = \frac{dv_0}{dx} \tag{4.25}$$

From this, the strain-displacement relational expression of the Bernoulli-Euler beam theory is as follows:

$$\left\{ \begin{array}{c} \varepsilon_{xx} \\ 2\varepsilon_{xy} \\ 2\varepsilon_{zx} \end{array} \right\} = \left(\begin{array}{ccc|c} \frac{d}{dx} & -y\frac{d^2}{dx^2} & -z\frac{d^2}{dx^2} & 0 \\ 0 & 0 & 0 & -z\frac{d}{dx} \\ 0 & 0 & 0 & y\frac{d}{dx} \end{array} \right) \left\{ \begin{array}{c} u_0 \\ v_0 \\ w_0 \\ \theta_x \end{array} \right\}$$

$$\varepsilon = L\hat{u} \tag{4.26}$$

The relational expression between stress and strain is as follows:

$$\left\{ \begin{array}{c} \sigma_{xx} \\ \sigma_{xy} \\ \sigma_{zx} \end{array} \right\} = \left(\begin{array}{ccc} E & 0 & 0 \\ 0 & G & 0 \\ 0 & 0 & G \end{array} \right) \left\{ \begin{array}{c} \varepsilon_{xx} \\ 2\varepsilon_{xy} \\ 2\varepsilon_{zx} \end{array} \right\}$$

$$\sigma = D\varepsilon \tag{4.27}$$

The strain energy $\mathcal{U}$ is expressed as follows:

$$\mathcal{U} = \frac{1}{2}\int_V \left\{\begin{array}{c} \sigma_{xx} \\ \sigma_{xy} \\ \sigma_{zx} \end{array}\right\}^{\mathrm{T}} \left\{\begin{array}{c} \varepsilon_{xx} \\ 2\varepsilon_{xy} \\ 2\varepsilon_{zx} \end{array}\right\} dV \tag{4.28}$$

The load acting on the inside of the member is $\boldsymbol{p} = \{p_x, p_y, p_z, m_x, m_y, m_z\}^{\mathrm{T}}$, and the load acting on the surface is $\boldsymbol{P} = \{\overline{P}_x, \overline{P}_y, \overline{P}_z, \overline{M}_x, \overline{M}_y, \overline{M}_z\}^{\mathrm{T}}$. The work $\mathcal{W}$ by load is given by the following equation.

$$\mathcal{W} = \int_0^\ell \boldsymbol{p}^{\mathrm{T}}\hat{\boldsymbol{u}}\,dx + \left[\overline{n}_x\overline{\boldsymbol{P}}^{\mathrm{T}}\hat{\boldsymbol{u}}\right]_0^\ell \tag{4.29}$$

Here, ℓ is the member length of the rigid frame member, and $\overline{n}_x$ is the direction cosine between the outward normal of the cross-section of the member end and the x-axis, $x = 0$ for $\overline{n}_x = -1$, $x = \ell$ for $\overline{n}_x = 1$.

From these, the total potential energy Π is defined by the following equation.

$$\Pi = \mathcal{U} - \mathcal{W} = \Pi(\hat{\boldsymbol{u}}) \tag{4.30}$$

4.4 Fundamentals of the Finite Element Method

4.4.1 Finite Element

When the area of the entire structure is subdivided, displacement and rotation can be expressed by simple functions in that area. Moreover, within this small region finite element, the shape of a complicated boundary can be expressed in a simple form. Finite elements are defined for each mechanical theory, such as solid members, plate and shell members, and rigid frame members, as in the previous chapter. In addition, various elements are considered depending on the nodal points and the shapes of the elements that constitute the elements.

4.4.2 Displacement Function and Shape Function

In the finite element method, interpolation is performed from the value at a specific point in the region (this is called a **nodal point**) to represent the displacement or rotation. For example, for a beam element of length ℓ with nodal points on both sides of the region, the displacement $u(x)$ is u_1, u_2 from displacements at nodal points 1 and 2 (node displacement), and is expressed as follows:

$$\begin{aligned} u(x) &= N_1(x)u_1 + N_2(x)u_2, \\ N_1(x) &= \frac{x}{\ell}, \quad N_2(x) = 1 - \frac{x}{\ell} \end{aligned} \tag{4.31}$$

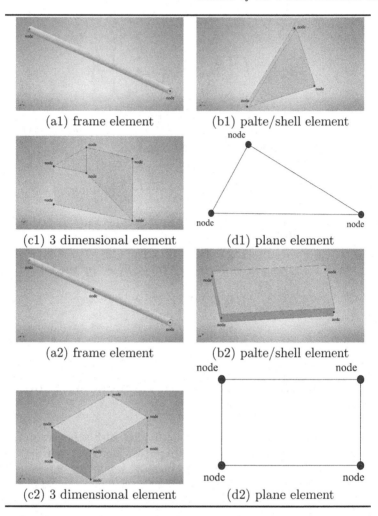

(a1) frame element   (b1) palte/shell element

(c1) 3 dimensional element   (d1) plane element

(a2) frame element   (b2) palte/shell element

(c2) 3 dimensional element   (d2) plane element

FIGURE 4.7
Type of finite element

Functions that can be applied to node displacements, such as $N_1(x)$ and $N_2(x)$ in the above equation, are called **Shape Function**. Also, the product of the node displacement and the shape function as shown in the above equation is called **Displacement Function**. Node displacement is a general term for multiple components such as displacement and rotation, and the number of components differs depending on the type of finite element.

- Three-dimensional beam element: 6 components of displacement u, v, w, rotation $\theta_x, \theta_y, \theta_z$

- Two-dimensional beam element: displacement u, v, rotation θ_z

- plate and shell elements: 6 components of displacement u, v, w, rotation $\theta_x, \theta_y, \theta_z$

- Three-dimensional solid element: 3 components of displacement u, v, w

- Two-dimensional solid element: 2 components of displacement u, v

These components are called **nodal degrees of freedom**.

4.4.3 Formulation of Stiffness Equation

Derivation of the general form of the equilibrium equation (rigidity equation) of elements consisting of n nodes. The displacement vector $\boldsymbol{u}$ in the element is expressed as follows from the displacement vector $\boldsymbol{d}_i$ and **Shape Function** of node i.

$$\boldsymbol{u} \simeq N_1 \boldsymbol{d}_1 + \cdots + N_n \boldsymbol{d}_n = \sum_{i=1}^{n} N_i \boldsymbol{d}_i \qquad (4.32)$$

Here, N_i is a matrix of shape functions for the node i. Vector $\boldsymbol{\varepsilon}$ consists of strain components and vector $\boldsymbol{\sigma}$ consists of stress components. From the relationship between strain and displacement and the relationship between stress and strain, it is expressed as follows from the node displacement.

$$\boldsymbol{\varepsilon} = \sum_{i=1}^{n} B_i \boldsymbol{d}_i, \quad \boldsymbol{\sigma} = D\boldsymbol{\varepsilon} \qquad (4.33)$$

Here, B_i is called the strain matrix, and D is called the elastic matrix. At this time, the area occupied by the object is V, the area on the surface of the object is S, the vector of force acting inside of the object is $\boldsymbol{p}$, and the vector of the surface force acting on the surface of the object is $\boldsymbol{q}$. The total potential energy Π is generally expressed as follows:

$$\Pi = \frac{1}{2} \sum_{i=1}^{n} \sum_{j=1}^{n} \boldsymbol{d}_i^{\mathrm{T}} \int_V B_i^{\mathrm{T}} D B_j dV \, \boldsymbol{d}_j - \sum_{i=1}^{n} \boldsymbol{d}_i^{\mathrm{T}} \boldsymbol{f}_i \qquad (4.34)$$

Here, $\bullet^{\mathrm{T}}$ represents the transpose of the vector $\bullet$.

$$\boldsymbol{f}_i = \int_V N_i^{\mathrm{T}} \boldsymbol{p} \, dV + \int_S N_i^{\mathrm{T}} \boldsymbol{q} \, dS$$

where $\boldsymbol{f}_i$ is the **equivalent nodal load vector**.
    This first variation is

$$\delta \Pi = \sum_{i=1}^{n} \delta \boldsymbol{d}_i^{\mathrm{T}} \left\{ \sum_{j=1}^{n} \int_V B_i^{\mathrm{T}} D B_j \, dV \, \boldsymbol{d}_j - \boldsymbol{f}_i \right\} \qquad (4.35)$$

The equation for the equilibrium state is obtained from $\delta\Pi = 0$ from the **principle of minimum total potential energy** and is as follows:

$$\sum_{j=1}^{n}[K_{ij}]\boldsymbol{d}_j = \boldsymbol{f}_i, \quad i = 1, \cdots, n \tag{4.36}$$

Here, $[K_{ij}]$ is obtained by the following equation.

$$[K_{ij}] = \int_V \boldsymbol{B}_i^{\mathrm{T}} D \boldsymbol{B}_j \ dV \tag{4.37}$$

Here, $[K_{ij}] \in K_e$ is called the **element stiffness matrix** related to the nodes i, j.

4.5 Formulation of Finite Elements

4.5.1 Plane Element

The nodal displacement vector, strain vector, and stress vector of the two-dimensional plane element are

$$\boldsymbol{d}_i = \left\{ \begin{array}{c} u_i \\ v_i \end{array} \right\}, \quad \boldsymbol{\varepsilon} = \left\{ \begin{array}{c} \varepsilon_{xx} \\ \varepsilon_{yy} \\ 2\varepsilon_{xy} \end{array} \right\}, \quad \boldsymbol{\sigma} = \left\{ \begin{array}{c} \sigma_{xx} \\ \sigma_{yy} \\ \sigma_{xy} \end{array} \right\}$$

Shape function matrix and strain matrix are

$$\boldsymbol{N}_i = \left(\begin{array}{cc} N_i & 0 \\ 0 & N_i \end{array} \right), \quad \boldsymbol{B}_i = \left(\begin{array}{cc} \frac{\partial N_i}{\partial x} & 0 \\ 0 & \frac{\partial N_i}{\partial y} \\ \frac{\partial N_i}{\partial y} & \frac{\partial N_i}{\partial x} \end{array} \right)$$

In the plane stress problem, the material composition matrix is following;

$$D = \frac{E}{1-\nu^2} \left(\begin{array}{ccc} 1 & \nu & 0 \\ \nu & 1 & 0 \\ 0 & 0 & (1-\nu)/2 \end{array} \right) \tag{4.38}$$

Next, in the plane strain problem,

$$D = \frac{Et}{(1+\nu)(1-2\nu)} \left(\begin{array}{ccc} 1-\nu & \nu & 0 \\ \nu & 1-\nu & 0 \\ 0 & 0 & (1-2\nu)/2 \end{array} \right) \tag{4.39}$$

is used. The element stiffness matrix for the nodes i, j is expressed as follows:

$$[K_{ij}^e] = \int_V \boldsymbol{B}_i^{\mathrm{T}} D \boldsymbol{B}_j dV, \quad i, j = 1, \cdots, n \tag{4.40}$$

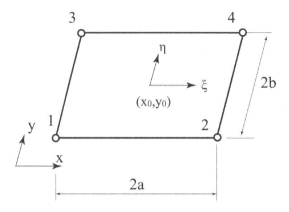

FIGURE 4.8
4-Node rectangular element

In the element stiffness matrix of plane stress,

$$[K_{ij}^e(\sigma)] = \frac{Et}{1-\nu^2}\left[\left(\begin{array}{cc} \frac{\partial N_i \partial N_j}{\partial x^2} & \frac{\partial N_i \partial N_j}{\partial x \partial y}\nu \\ \frac{\partial N_i \partial N_j}{\partial x \partial y}\nu & \frac{\partial N_i \partial N_j}{\partial y^2} \end{array}\right) + \frac{1-\nu}{2}\left(\begin{array}{cc} \frac{\partial N_i \partial N_j}{\partial y^2} & \frac{\partial N_i \partial N_j}{\partial x \partial y} \\ \frac{\partial N_i \partial N_j}{\partial x \partial y} & \frac{\partial N_i \partial N_j}{\partial x^2} \end{array}\right)\right]$$

Next, in the element stiffness matrix of plane strain,

$$[K_{ij}^e(\varepsilon)] = \frac{Et}{(1+\nu)(1-2\nu)}\left[\left(\begin{array}{cc} \frac{\partial N_i \partial N_j}{\partial x^2}(1-\nu) & \frac{\partial N_i \partial N_j}{\partial x \partial y}\nu \\ \frac{\partial N_i \partial N_j}{\partial x \partial y}\nu & \frac{\partial N_i \partial N_j}{\partial y^2}(1-\nu) \end{array}\right)\right.$$
$$\left. +\frac{1-2\nu}{2}\left(\begin{array}{cc} \frac{\partial N_i \partial N_j}{\partial y^2} & \frac{\partial N_i \partial N_j}{\partial x \partial y} \\ \frac{\partial N_i \partial N_j}{\partial x \partial y} & \frac{\partial N_i \partial N_j}{\partial x^2} \end{array}\right)\right]$$

The stiffness matrix of the plane stress state is $\lambda + 2G$ and λ, which are $E/(1-\nu^2)$ and $\nu E/(1-\nu^2)$, respectively. The load vector $\boldsymbol{f}_i$ is as follows:

$$\boldsymbol{f}_i = \int_V \left\{ \begin{array}{c} N_i p_{xi} \\ N_i p_{yi} \end{array} \right\} dV + \int_S \left\{ \begin{array}{c} N_i q_{xi} \\ N_i q_{yi} \end{array} \right\} dS$$

4.5.2 4-Node Rectangular Element

Induce the shape function of a 4 nodes rectangular element of size 2a × 2b, such as **Fig. 4.8**. To simplify the derivation, consider the normal coordinate system (ξ, η), which takes the value of ± 1 at each node. The relationship between this coordinate system and the (x, y) coordinate system is as follows:

$$\xi = \frac{x - x_0}{a}, \quad \eta = \frac{y - y_0}{b}$$

When finding the stiffness matrix, it is necessary to perform differential operations on (x, y) of these shape functions. Therefore, the differential relationship between (x, y) coordinate and (ξ, η) coordinate is shown below.

$$\frac{\partial}{\partial x} = \frac{\partial \xi}{\partial x} \frac{\partial}{\partial \xi} = \frac{1}{a} \frac{\partial}{\partial \xi}, \quad \frac{\partial}{\partial y} = \frac{\partial \eta}{\partial y} \frac{\partial}{\partial \eta} = \frac{1}{b} \frac{\partial}{\partial \eta}$$

Also, the integral is expressed as

$$V = \int_V dV = ab \int\int_{-1}^{1} d\xi \, d\eta. \tag{4.41}$$

Since there are four nodes, each displacement u_i, v_i can be expressed using four undetermined coefficients, so u_i is expressed as follows:

$$u_i = a_1 + a_2 \xi + a_3 \eta + a_4 \xi \eta, \quad i = 1, \cdots, 4$$

Here, $a_i (i = 1, 2, 3, 4)$ is an undetermined coefficient.

$$\begin{Bmatrix} u_1 \\ u_2 \\ u_3 \\ u_4 \end{Bmatrix} = \begin{pmatrix} 1 & -1 & -1 & 1 \\ 1 & 1 & -1 & -1 \\ 1 & -1 & 1 & -1 \\ 1 & 1 & 1 & 1 \end{pmatrix} \begin{Bmatrix} a_1 \\ a_2 \\ a_3 \\ a_4 \end{Bmatrix}$$

Solving this system of equations with respect to the undetermined coefficients

$$\begin{Bmatrix} a_1 \\ a_2 \\ a_3 \\ a_4 \end{Bmatrix} = \frac{1}{4} \begin{pmatrix} 1 & 1 & 1 & 1 \\ -1 & 1 & -1 & 1 \\ -1 & -1 & 1 & 1 \\ 1 & -1 & -1 & 1 \end{pmatrix} \begin{Bmatrix} u_1 \\ u_2 \\ u_3 \\ u_4 \end{Bmatrix}$$

When the displacement v is also derived in the same way, the displacements u, v are expressed by the node displacements as follows:

$$u = \sum_{i=1}^{4} N_i u_i, \quad v = \sum_{i=1}^{4} N_i v_i$$

Here, the shape function is given by the following equation. The normalized 4-node shape function

$$N_j = \frac{1}{4}(1 + \xi_j \xi)(1 + \eta_j \eta), \quad j = 1, \cdots, 4 \tag{4.42}$$

is given. Here, from the normal coordinate points of the nodes $(\xi_1, \eta_1) = (-1, -1)$, $(\xi_2, \eta_2) = (1, -1)$, $(\xi_3, \eta_3) = (-1, 1)$, $(\xi_4, \eta_4) = (1, 1)$ are given.

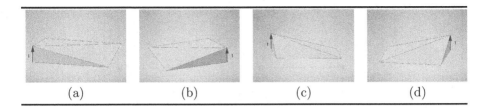

(a)   (b)   (c)   (d)

FIGURE 4.9
4-Node rectangular element shape function

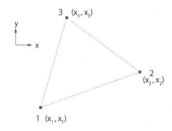

FIGURE 4.10
3-Node triangle element

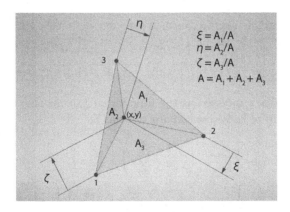

FIGURE 4.11
Triangle coordinates (area coordinates)

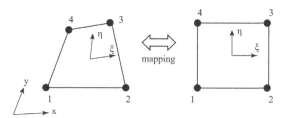

FIGURE 4.12
4-Node isoparametric quadrilateral element

4.5.3   4-Node Isoparametric Quadrilateral Element

Next, the shape function of a 4-node quadrilateral element of arbitrary shape, as shown on the left of **Fig. 4.12**, and the method of deriving the stiffness matrix are shown. If such an element of arbitrary shape is also expressed in the (x, y) coordinate system, the formula will become complicated. To avoid this situation, the value in the range of ± 1 can be considered under the coordinate system (ξ, η). Here, if assuming, the coordinate ξ takes a value of -1 on side 13 and $+1$ on side 24. And the coordinate η takes a value of -1 on side 12 and $+1$ on side 34. The value of (x, y) at the (ξ, η) point in the element can be expressed as follows from the value of (x, y) at the four nodes.

$$x = \sum_{i=1}^{4} \widetilde{N}_i x_i, \quad v = \sum_{i=1}^{4} \widetilde{N}_i y_i$$

Here, N_i is a function that expresses the shape of the element, and is given by the following equation.

$$
\begin{aligned}
\widetilde{N}_1 &= \frac{1}{4}(1 - \xi)(1 - \eta), \quad \widetilde{N}_2 = \frac{1}{4}(1 + \xi)(1 - \eta), \\
\widetilde{N}_3 &= \frac{1}{4}(1 - \xi)(1 + \eta), \quad \widetilde{N}_4 = \frac{1}{4}(1 + \xi)(1 + \eta)
\end{aligned}
$$

That is, x, y can be expressed as the function of ξ, η. At this time, the differential relationship of these coordinate systems is

$$
\left\{ \begin{array}{c} \frac{\partial}{\partial x} \\ \frac{\partial}{\partial y} \end{array} \right\} = \left(\begin{array}{cc} \frac{\partial x}{\partial \xi} & \frac{\partial y}{\partial \xi} \\ \frac{\partial x}{\partial \eta} & \frac{\partial y}{\partial \eta} \end{array} \right) \left\{ \begin{array}{c} \frac{\partial}{\partial x} \\ \frac{\partial}{\partial y} \end{array} \right\}
$$

$$
\left\{ \begin{array}{c} \frac{\partial}{\partial \xi} \\ \frac{\partial}{\partial \eta} \end{array} \right\} = \frac{1}{\det J} \left(\begin{array}{cc} \frac{\partial y}{\partial \eta} & -\frac{\partial y}{\partial \xi} \\ -\frac{\partial x}{\partial \eta} & \frac{\partial x}{\partial \xi} \end{array} \right) \left\{ \begin{array}{c} \frac{\partial}{\partial \xi} \\ \frac{\partial}{\partial \eta} \end{array} \right\}
$$

Here,

$$
\det J = \left| \begin{array}{cc} \frac{\partial x}{\partial \xi} & \frac{\partial y}{\partial \xi} \\ \frac{\partial x}{\partial \eta} & \frac{\partial y}{\partial \eta} \end{array} \right|
$$

The integral is expressed as follows:

$$V = \int_V dV = \int\int_{-1}^{1} \det J \, d\xi d\eta$$

Considering such a (ξ, η) coordinate system, a quadrilateral element of arbitrary shape can be thought of as a rectangular element, and with $\det J = ab$, the volume integral V is the same as Eq.(4.41). The displacement u, v in the element can be expressed as follows from the shape function expressed in (ξ, η) coordinates.

$$u = \sum_{i=1}^{4} N_i u_i, \quad v = \sum_{i=1}^{4} N_i v_i$$

At this time, the shape functionshape function N_i is the same as the shape function of the rectangular element.

$$N_1 = \frac{1}{4}(1 - \xi)(1 - \eta), \quad N_2 = \frac{1}{4}(1 + \xi)(1 - \eta)$$
$$N_3 = \frac{1}{4}(1 - \xi)(1 + \eta), \quad N_4 = \frac{1}{4}(1 + \xi)(1 + \eta)$$

Also, the function $\widetilde{N}_i$, which represents the coordinate values in the element, is the same function as N_i. In this way, an element that uses the same function for expressing the displacement in the element and the function for expressing the coordinate value (element shape) in the element is called an **isoparametric element**.

4.5.4   3D Solid Element

The nodal displacement vector, strain vector, stress vector, shape function matrix, strain matrix, and elastic matrix of 3D solid elements are as follows:

$$d_i = \left\{ \begin{array}{c} u_i \\ v_i \\ w_i \end{array} \right\}, \quad \varepsilon = \left\{ \begin{array}{c} \varepsilon_{xx} \\ \varepsilon_{yy} \\ \varepsilon_{zz} \\ 2\varepsilon_{xy} \\ 2\varepsilon_{yz} \\ 2\varepsilon_{zx} \end{array} \right\}, \quad \sigma = \left\{ \begin{array}{c} \sigma_{xx} \\ \sigma_{yy} \\ \sigma_{zz} \\ \sigma_{xy} \\ \sigma_{yz} \\ \sigma_{zx} \end{array} \right\} \tag{4.43}$$

$$N_i = \begin{pmatrix} N_i & 0 & 0 \\ 0 & N_i & 0 \\ 0 & 0 & N_i \end{pmatrix}, \quad B_i = \begin{pmatrix} \frac{\partial N_i}{\partial x} & 0 & 0 \\ 0 & \frac{\partial N_i}{\partial y} & 0 \\ 0 & 0 & \frac{\partial N_i}{\partial z} \\ \frac{\partial N_i}{\partial y} & \frac{\partial N_i}{\partial x} & 0 \\ 0 & \frac{\partial N_i}{\partial z} & \frac{\partial N_i}{\partial y} \\ \frac{\partial N_i}{\partial z} & 0 & \frac{\partial N_i}{\partial x} \end{pmatrix} \tag{4.44}$$

$$D = \begin{pmatrix} \lambda + 2G & \lambda & \lambda & 0 & 0 & 0 \\ \lambda & \lambda + 2G & \lambda & 0 & 0 & 0 \\ \lambda & \lambda & \lambda + 2G & 0 & 0 & 0 \\ \hline 0 & 0 & 0 & G & 0 & 0 \\ 0 & 0 & 0 & 0 & G & 0 \\ 0 & 0 & 0 & 0 & 0 & G \end{pmatrix} \qquad (4.45)$$

From these, the stiffness matrix is expressed as follows:

$$K^e = \int_V \begin{pmatrix} k_{11}^e & k_{12}^e & k_{13}^e \\ k_{21}^e & k_{22}^e & k_{23}^e \\ k_{31}^e & k_{32}^e & k_{33}^e \end{pmatrix} dx\, dy \qquad (4.46)$$

where k_{ij}^e

$$k_{11}^e = (\lambda + 2G)\frac{\partial N_i}{\partial x}\frac{\partial N_j}{\partial x} + G\left(\frac{\partial N_i}{\partial y}\frac{\partial N_j}{\partial y} + \frac{\partial N_i}{\partial z}\frac{\partial N_j}{\partial z}\right) \qquad (4.47)$$

$$k_{12}^e = \lambda\frac{\partial N_i}{\partial x}\frac{\partial N_j}{\partial y} + G\frac{\partial N_i}{\partial y}\frac{\partial N_j}{\partial x} \qquad (4.48)$$

$$k_{13}^e = \lambda\frac{\partial N_i}{\partial x}\frac{\partial N_j}{\partial z} + G\frac{\partial N_i}{\partial z}\frac{\partial N_j}{\partial x} \qquad (4.49)$$

$$k_{21}^e = \lambda\frac{\partial N_i}{\partial y}\frac{\partial N_j}{\partial x} + G\frac{\partial N_i}{\partial x}\frac{\partial N_j}{\partial y} \qquad (4.50)$$

$$k_{22}^e = (\lambda + 2G)\frac{\partial N_i}{\partial y}\frac{\partial N_j}{\partial y} + G\left(\frac{\partial N_i}{\partial z}\frac{\partial N_j}{\partial z} + \frac{\partial N_i}{\partial x}\frac{\partial N_j}{\partial x}\right) \qquad (4.51)$$

$$k_{23}^e = \lambda\frac{\partial N_i}{\partial y}\frac{\partial N_j}{\partial z} + G\frac{\partial N_i}{\partial z}\frac{\partial N_j}{\partial y} \qquad (4.52)$$

$$k_{31}^e = \lambda\frac{\partial N_i}{\partial z}\frac{\partial N_j}{\partial x} + G\frac{\partial N_i}{\partial x}\frac{\partial N_j}{\partial z} \qquad (4.53)$$

$$k_{32}^e = \lambda\frac{\partial N_i}{\partial z}\frac{\partial N_j}{\partial y} + G\frac{\partial N_i}{\partial y}\frac{\partial N_j}{\partial z} \qquad (4.54)$$

$$k_{33}^e = (\lambda + 2G)\frac{\partial N_i}{\partial z}\frac{\partial N_j}{\partial z} + G\left(\frac{\partial N_i}{\partial x}\frac{\partial N_j}{\partial x} + \frac{\partial N_i}{\partial y}\frac{\partial N_j}{\partial y}\right) \qquad (4.55)$$

The load vector $\boldsymbol{f}_i$ looks like this,

$$f_i = \int_V \left\{ \begin{array}{c} N_i p_{xi} \\ N_i p_{yi} \\ N_i p_{zi} \end{array} \right\} dV + \int_S \left\{ \begin{array}{c} N_i Q_{xi} \\ N_i Q_{yi} \\ N_i Q_{zi} \end{array} \right\} dS \qquad (4.56)$$

4.5.4.1   8-Node Rectangular Parallelepiped Element

As shown in **Fig. 4.13,** the shape function of the 8-node rectangular parallelepiped element with the size of $2abc$ can be derived. To simplify the derivation, consider a normal coordinate system (ξ, η, ζ) that takes the value of ± 1

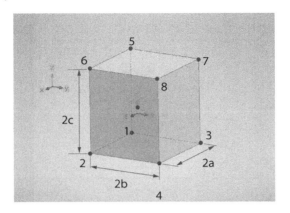

FIGURE 4.13
8-Node rectangular parallelepiped element

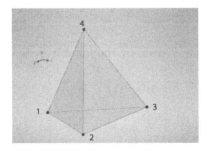

FIGURE 4.14
4-node tetrahedral element

at each node. The relationship between this coordinate system and (x, y, z) coordinate system is as follows:

$$\xi = \frac{x - x_G}{a}, \quad \eta = \frac{y - y_G}{b}, \quad \zeta = \frac{z - z_G}{c} \tag{4.57}$$

When finding the stiffness matrix, differential operations related to x, y, z of these shape functions are required. Therefore, the differential relationship between (x, y, z) coordinates and (ξ, η, ζ) coordinates is shown below.

$$\frac{\partial}{\partial x} = \frac{\partial \xi}{\partial x} \frac{\partial}{\partial \xi} = \frac{1}{a} \frac{\partial}{\partial \xi}, \quad \frac{\partial}{\partial y} = \frac{\partial \eta}{\partial y} \frac{\partial}{\partial \eta} = \frac{1}{b} \frac{\partial}{\partial \eta}, \quad \frac{\partial}{\partial z} = \frac{\partial \zeta}{\partial z} \frac{\partial}{\partial \zeta} = \frac{1}{c} \frac{\partial}{\partial \zeta} \tag{4.58}$$

The integral is expressed as follows:

$$V = ab \int_{-1}^{1} \int_{-1}^{1} \int_{-1}^{1} d\xi \, d\eta \, d\zeta \tag{4.59}$$

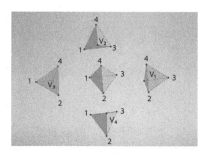

FIGURE 4.15
Volume coordinates

The displacements u, v, w are expressed as node displacements as follows:

$$u = \sum_{i=1}^{8} N_i u_i, \quad v = \sum_{i=1}^{8} N_i v_i, \quad w = \sum_{i=1}^{8} N_i w_i \qquad (4.60)$$

Here, the shape function is given by the following equation.

$$
\begin{aligned}
N_1 &= \frac{1}{8}(1-\xi)(1-\eta)(1-\zeta), & N_2 &= \frac{1}{8}(1+\xi)(1-\eta)(1-\zeta) \\
N_3 &= \frac{1}{8}(1-\xi)(1+\eta)(1-\zeta), & N_4 &= \frac{1}{8}(1+\xi)(1+\eta)(1-\zeta) \\
N_5 &= \frac{1}{8}(1-\xi)(1+\eta)(1-\zeta), & N_6 &= \frac{1}{8}(1+\xi)(1+\eta)(1-\zeta) \\
N_7 &= \frac{1}{8}(1-\xi)(1+\eta)(1+\zeta), & N_8 &= \frac{1}{8}(1+\xi)(1+\eta)(1+\zeta)
\end{aligned}
$$

4.5.4.2   4-Node tetrahedral element

Again, instead of the (x, y, z) coordinates, we use a coordinate system called tetrahedral coordinates or volume coordinates. This coordinate system can be created by connecting points (x, y, z) in the element to each vertex of the tetrahedron as shown in **Fig. 4.15**. The ratio of volume V_1, V_2, V_3, V_4 of four tetrahedra and volume V of element

$$\xi = \frac{V_1}{V}, \quad \eta = \frac{V_2}{V}, \quad \zeta = \frac{V_3}{V}, \quad \omega = \frac{V_4}{V} \qquad (4.61)$$

expressed by the (ξ, η, ζ). However, there is a relationship of $\xi + \eta + \zeta\omega = 1$ between these four coordinate values. The volume of each tetrahedron is expressed as follows:

$$6V_1 = \begin{vmatrix} x_2 & x_3 & x_4 \\ y_2 & y_3 & y_4 \\ z_2 & z_3 & z_4 \end{vmatrix}$$

$$+x \begin{vmatrix} y_2 & y_3 & y_4 \\ z_2 & z_3 & z_4 \\ 1 & 1 & 1 \end{vmatrix} + y \begin{vmatrix} z_2 & z_3 & z_4 \\ 1 & 1 & 1 \\ x_2 & x_3 & x_4 \end{vmatrix} + z \begin{vmatrix} 1 & 1 & 1 \\ x_2 & x_3 & x_4 \\ y_2 & y_3 & y_4 \end{vmatrix}$$

$$6V_2 = \begin{vmatrix} x_3 & x_4 & x_1 \\ y_3 & y_4 & y_1 \\ z_3 & z_4 & z_1 \end{vmatrix}$$

$$+x \begin{vmatrix} y_3 & y_4 & y_1 \\ z_3 & z_4 & z_1 \\ 1 & 1 & 1 \end{vmatrix} + y \begin{vmatrix} z_3 & z_4 & z_1 \\ 1 & 1 & 1 \\ x_3 & x_4 & x_1 \end{vmatrix} + z \begin{vmatrix} 1 & 1 & 1 \\ x_3 & x_4 & x_1 \\ y_3 & y_4 & y_1 \end{vmatrix}$$

$$6V_3 = \begin{vmatrix} x_4 & x_1 & x_2 \\ y_4 & y_1 & y_2 \\ z_4 & z_1 & z_2 \end{vmatrix}$$

$$+x \begin{vmatrix} y_4 & y_1 & y_2 \\ z_4 & z_1 & z_2 \\ 1 & 1 & 1 \end{vmatrix} + y \begin{vmatrix} z_4 & z_1 & z_2 \\ 1 & 1 & 1 \\ x_4 & x_1 & x_2 \end{vmatrix} + z \begin{vmatrix} 1 & 1 & 1 \\ x_4 & x_1 & x_2 \\ y_4 & y_1 & y_2 \end{vmatrix}$$

$$6V_4 = \begin{vmatrix} x_1 & x_2 & x_3 \\ y_1 & y_2 & y_3 \\ z_1 & z_2 & z_3 \end{vmatrix}$$

$$+x \begin{vmatrix} y_1 & y_2 & y_3 \\ z_1 & z_2 & z_3 \\ 1 & 1 & 1 \end{vmatrix} + y \begin{vmatrix} z_1 & z_2 & z_3 \\ 1 & 1 & 1 \\ x_1 & x_2 & x_3 \end{vmatrix} + z \begin{vmatrix} 1 & 1 & 1 \\ x_1 & x_2 & x_3 \\ y_1 & y_2 & y_3 \end{vmatrix}$$

$$V = \begin{vmatrix} 1 & 1 & 1 & 1 \\ x_1 & x_2 & x_3 & x_4 \\ y_1 & y_2 & y_3 & y_4 \\ z_1 & z_2 & z_3 & z_4 \end{vmatrix}$$

The differential relationship between the (x, y, z) coordinate system and the (ξ, η, ζ) coordinate system is

$$\frac{\partial}{\partial x} = \frac{\partial \xi}{\partial x} \frac{\partial}{\partial \xi} + \frac{\partial \eta}{\partial x} \frac{\partial}{\partial \eta} + \frac{\partial \zeta}{\partial x} \frac{\partial}{\partial \zeta} + \frac{\partial \omega}{\partial x} \frac{\partial}{\partial \omega}$$

$$= \frac{1}{6V} \begin{vmatrix} y_2 & y_3 & y_4 \\ z_2 & z_3 & z_4 \\ 1 & 1 & 1 \end{vmatrix} \frac{\partial}{\partial \xi} + \frac{1}{6V} \begin{vmatrix} y_3 & y_4 & y_1 \\ z_3 & z_4 & z_1 \\ 1 & 1 & 1 \end{vmatrix} \frac{\partial}{\partial \eta}$$

$$+ \frac{1}{6V} \begin{vmatrix} y_4 & y_1 & y_2 \\ z_4 & z_1 & z_2 \\ 1 & 1 & 1 \end{vmatrix} \frac{\partial}{\partial \zeta} + \frac{1}{6V} \begin{vmatrix} y_1 & y_2 & y_3 \\ z_1 & z_2 & z_3 \\ 1 & 1 & 1 \end{vmatrix} \frac{\partial}{\partial \omega}$$

$$\frac{\partial}{\partial y} = \frac{\partial \xi}{\partial y}\frac{\partial}{\partial \xi} + \frac{\partial \eta}{\partial y}\frac{\partial}{\partial \eta} + \frac{\partial \zeta}{\partial y}\frac{\partial}{\partial \zeta} + \frac{\partial \omega}{\partial y}\frac{\partial}{\partial \omega}$$

$$= \frac{1}{6V}\begin{vmatrix} z_2 & z_3 & z_4 \\ 1 & 1 & 1 \\ x_2 & x_3 & x_4 \end{vmatrix}\frac{\partial}{\partial \xi} + \frac{1}{6V}\begin{vmatrix} z_3 & z_4 & z_1 \\ 1 & 1 & 1 \\ x_3 & x_4 & x_1 \end{vmatrix}\frac{\partial}{\partial \eta}$$

$$+ \frac{1}{6V}\begin{vmatrix} z_4 & z_1 & z_2 \\ 1 & 1 & 1 \\ x_4 & x_1 & x_2 \end{vmatrix}\frac{\partial}{\partial \zeta} + \frac{1}{6V}\begin{vmatrix} z_1 & z_2 & z_3 \\ 1 & 1 & 1 \\ x_1 & x_2 & x_3 \end{vmatrix}\frac{\partial}{\partial \omega}$$

$$\frac{\partial}{\partial z} = \frac{\partial \xi}{\partial z}\frac{\partial}{\partial \xi} + \frac{\partial \eta}{\partial z}\frac{\partial}{\partial \eta} + \frac{\partial \zeta}{\partial z}\frac{\partial}{\partial \zeta} + \frac{\partial \omega}{\partial z}\frac{\partial}{\partial \omega}$$

$$= \frac{1}{6V}\begin{vmatrix} 1 & 1 & 1 \\ x_2 & x_3 & x_4 \\ y_2 & y_3 & y_4 \end{vmatrix}\frac{\partial}{\partial \xi} + \frac{1}{6V}\begin{vmatrix} 1 & 1 & 1 \\ x_3 & x_4 & x_1 \\ y_3 & y_4 & y_1 \end{vmatrix}\frac{\partial}{\partial \eta}$$

$$+ \frac{1}{6V}\begin{vmatrix} 1 & 1 & 1 \\ x_4 & x_1 & x_2 \\ y_4 & y_1 & y_2 \end{vmatrix}\frac{\partial}{\partial \zeta} + \frac{1}{6V}\begin{vmatrix} 1 & 1 & 1 \\ x_1 & x_2 & x_3 \\ y_1 & y_2 & y_3 \end{vmatrix}\frac{\partial}{\partial \omega}$$

The integral is expressed as follows:

$$\int_V dV = 6V \int_0^1 \int_0^{1-\zeta} \int_0^{1-\eta-\zeta} d\xi d\eta d\zeta \tag{4.62}$$

The following general formula is derived for the exponential of tetrahedral coordinates.

$$\int_V \xi^i \eta^j \zeta^k \omega^\ell dV = \frac{i! \cdot j! \cdot k! \cdot \ell!}{(i+j+k+\ell+3)!}6V \tag{4.63}$$

The displacement in the element is expressed as follows from the four undetermined coefficients.

$$u = a_1 + a_2\xi + a_3\eta + a_4\zeta \tag{4.64}$$

At node 1, the tetrahedral coordinates are $(\xi, \eta, \zeta, \omega) = (1,0,0,0)$, at node 2, the coordinates become $(0,1,0,0)$, at node 3, the coordinates become $(0,0,1,0)$, and at node 4, the coordinates become $(0,0,0,1)$. The undetermined coefficient is expressed as follows from the displacement at each node.

$$a_1 = u_4, \quad a_2 = u_1 - u_4, \quad a_3 = u_2 - u_4, \quad a_4 = u_3 - u_4 \tag{4.65}$$

From this, when the undetermined coefficient is eliminated, the displacement in the element is expressed as follows:

$$\begin{aligned} u &= \xi u_1 + \eta u_2 + \zeta u_3 + \omega u_4, \\ v &= \xi v_1 + \eta v_2 + \zeta v_3 + \omega v_4, \\ w &= \xi w_1 + \eta w_2 + \zeta w_3 + \omega w_4 \end{aligned}$$

Therefore, the shape function is as follows:

$$N_1 = \xi, \quad N_2 = \eta, \quad N_3 = \zeta, \quad N_4 = \omega \tag{4.66}$$

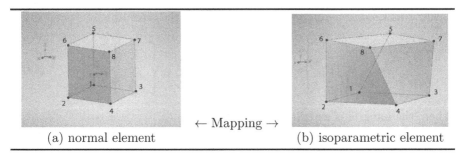

← Mapping →

(a) normal element　　　(b) isoparametric element

FIGURE 4.16
8-node isoparametric hexahedral element

4.5.4.3　8-Node Isoparametric Hexahedral Element

Next, the shape function of the 8-node hexahedron element of arbitrary shape and the method of deriving the stiffness matrix are shown on the left of **Fig. 4.16**. If such an element of arbitrary shape is also expressed in the (x, y, z) coordinate system, the formula will become complicated. To avoid this situation, the value in the range of ± 1 can be considered under the coordinate system (ξ, η, ζ). Here, the ξ coordinates take a value of -1 on surface 1357 and $+1$ on surface 2468. The η coordinates take a value of -1 on surface 1256 and $+1$ on surface 3478. The ζ coordinates are -1 on surface 1234 and $+1$ on surface 5678. From the value of (x, y, z) at the eight nodes, the value of (x, y, z) at the point (ξ, η, ζ) in the element can be expressed as follows:

$$
\begin{aligned}
x &= \widetilde{N_1}x_1 + \widetilde{N_2}x_2 + \widetilde{N_3}x_3 + \widetilde{N_4}x_4 + \widetilde{N_5}x_5 + \widetilde{N_6}x_6 + \widetilde{N_7}x_7 + \widetilde{N_8}x_8 \\
y &= \widetilde{N_1}y_1 + \widetilde{N_2}y_2 + \widetilde{N_3}y_3 + \widetilde{N_4}y_4 + \widetilde{N_5}y_5 + \widetilde{N_6}y_6 + \widetilde{N_7}y_7 + \widetilde{N_8}y_8 \\
z &= \widetilde{N_1}z_1 + \widetilde{N_2}z_2 + \widetilde{N_3}z_3 + \widetilde{N_4}z_4 + \widetilde{N_5}z_5 + \widetilde{N_6}z_6 + \widetilde{N_7}z_7 + \widetilde{N_8}z_8
\end{aligned}
$$

Here, N is a function that expresses the shape of the element and is given by the following equation.

$$
\begin{aligned}
\widetilde{N_1} &= \frac{1}{8}(1 - \xi)(1 - \eta)(1 - \zeta), & \widetilde{N_2} &= \frac{1}{8}(1 + \xi)(1 - \eta)(1 - \zeta) \\
\widetilde{N_3} &= \frac{1}{8}(1 - \xi)(1 + \eta)(1 - \zeta), & \widetilde{N_4} &= \frac{1}{8}(1 + \xi)(1 + \eta)(1 - \zeta) \\
\widetilde{N_5} &= \frac{1}{8}(1 - \xi)(1 + \eta)(1 - \zeta), & \widetilde{N_6} &= \frac{1}{8}(1 + \xi)(1 + \eta)(1 - \zeta) \\
\widetilde{N_7} &= \frac{1}{8}(1 - \xi)(1 + \eta)(1 + \zeta), & \widetilde{N_8} &= \frac{1}{8}(1 + \xi)(1 + \eta)(1 + \zeta)
\end{aligned}
$$

That is, x, y, z is a function of ξ, η, ζ.

At this time, the differential relationship of these coordinate systems is

$$
\left\{
\begin{array}{c}
\frac{\partial}{\partial x} \\
\frac{\partial}{\partial y} \\
\frac{\partial}{\partial z}
\end{array}
\right\}
=
\left(
\begin{array}{ccc}
\frac{\partial x}{\partial \xi} & \frac{\partial y}{\partial \xi} & \frac{\partial z}{\partial \xi} \\
\frac{\partial x}{\partial \eta} & \frac{\partial y}{\partial \eta} & \frac{\partial z}{\partial \eta} \\
\frac{\partial x}{\partial \zeta} & \frac{\partial y}{\partial \zeta} & \frac{\partial z}{\partial \zeta}
\end{array}
\right)^{-1}
\left\{
\begin{array}{c}
\frac{\partial}{\partial \xi} \\
\frac{\partial}{\partial \eta} \\
\frac{\partial}{\partial \zeta}
\end{array}
\right\}
\tag{4.67}
$$

Then, the integral is expressed as follows:

$$
\int_V dV = \int_{-1}^{1} \int_{-1}^{1} \int_{-1}^{1} \det J \, d\xi d\eta d\zeta
\tag{4.68}
$$

Here,

$$
\det J =
\begin{vmatrix}
\frac{\partial x}{\partial \xi} & \frac{\partial y}{\partial \xi} & \frac{\partial z}{\partial \xi} \\
\frac{\partial x}{\partial \eta} & \frac{\partial y}{\partial \eta} & \frac{\partial z}{\partial \eta} \\
\frac{\partial x}{\partial \zeta} & \frac{\partial y}{\partial \zeta} & \frac{\partial z}{\partial \zeta}
\end{vmatrix}
\tag{4.69}
$$

Considering such a (ξ, η, ζ) coordinate system, a hexahedral element of arbitrary shape can be considered in the same way as a rectangular parallelepiped element. The displacements u, v, w in the element can be expressed as follows from the shape function expressed in (ξ, η, ζ) coordinates.

$$
\begin{aligned}
u &= N_1 u_1 + N_2 u_2 + N_3 u_3 + N_4 u_4 + N_5 u_5 + N_6 u_6 + N_7 u_7 + N_8 u_8 \\
v &= N_1 v_1 + N_2 v_2 + N_3 v_3 + N_4 v_4 + N_5 v_5 + N_6 v_6 + N_7 v_7 + N_8 v_8 \\
w &= N_1 w_1 + N_2 w_2 + N_3 w_3 + N_4 w_4 + N_5 w_5 + N_6 w_6 + N_7 w_7 + N_8 w_8
\end{aligned}
$$

This shape function is given by the following equation and is the same as the shape function of the rectangular element.

$$
\begin{aligned}
N_1 &= \tfrac{1}{8}(1-\xi)(1-\eta)(1-\zeta) , & N_2 &= \tfrac{1}{8}(1+\xi)(1-\eta)(1-\zeta) \\
N_3 &= \tfrac{1}{8}(1-\xi)(1+\eta)(1-\zeta) , & N_4 &= \tfrac{1}{8}(1+\xi)(1+\eta)(1-\zeta) \\
N_5 &= \tfrac{1}{8}(1-\xi)(1+\eta)(1-\zeta) , & N_6 &= \tfrac{1}{8}(1+\xi)(1+\eta)(1-\zeta) \\
N_7 &= \tfrac{1}{8}(1-\xi)(1+\eta)(1+\zeta) , & N_8 &= \tfrac{1}{8}(1+\xi)(1+\eta)(1+\zeta)
\end{aligned}
$$

Also, the function $\widetilde{N}_i$, which represents the coordinate values in the element, is the same function as N_i. In this way, an element that uses the same function for expressing the displacement in the element and the function for expressing the coordinate value (element shape) in the element is called the **isoparametric element**.

4.5.5   Plate and Shell Element

The nodal displacement vector, strain vector, stress vector, shape function matrix , strain matrix and elastic matrix of plate and shell elements are as

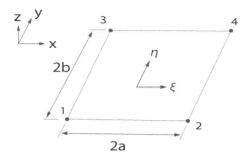

FIGURE 4.17
4-Node rectangular element

follows:

$$
\mathbf{d}_i = \left\{ \begin{array}{c} u_i \\ v_i \\ w_i \\ \hline \theta_{xi} \\ \theta_{yi} \\ \theta_{zi} \end{array} \right\}, \quad \boldsymbol{\varepsilon} = \left\{ \begin{array}{c} \varepsilon_{xx} \\ \varepsilon_{yy} \\ 2\varepsilon_{xy} \\ 2\varepsilon_{yz} \\ 2\varepsilon_{zx} \end{array} \right\}, \quad \boldsymbol{\sigma} = \left\{ \begin{array}{c} \sigma_{xx} \\ \sigma_{yy} \\ \sigma_{xy} \\ \sigma_{yz} \\ \sigma_{zx} \end{array} \right\} \tag{4.70}
$$

$$
N_i = \left(\begin{array}{ccc|ccc} N_i & 0 & 0 & 0 & zN_i & 0 \\ 0 & N_i & 0 & -zN_i & 0 & 0 \\ 0 & 0 & N_i & 0 & 0 & 0 \end{array} \right) \tag{4.71}
$$

$$
B_i = \left(\begin{array}{ccc|ccc} \frac{\partial N_i}{\partial x} & 0 & 0 & 0 & z\frac{\partial N_i}{\partial x} & 0 \\ 0 & \frac{\partial N_i}{\partial y} & 0 & -z\frac{\partial N_i}{\partial y} & 0 & 0 \\ \frac{\partial N_i}{\partial y} & \frac{\partial N_i}{\partial x} & 0 & -z\frac{\partial N_i}{\partial x} & z\frac{\partial N_i}{\partial y} & 0 \\ 0 & 0 & \frac{\partial N_i}{\partial y} & -N_i & 0 & 0 \\ 0 & 0 & \frac{\partial N_i}{\partial x} & 0 & N_i & 0 \end{array} \right) \tag{4.72}
$$

$$
D = \left(\begin{array}{ccccc} \frac{E}{1-\nu^2} & \frac{\nu E}{1-\nu^2} & 0 & 0 & 0 \\ \frac{\nu E}{1-\nu^2} & \frac{E}{1-\nu^2} & 0 & 0 & 0 \\ 0 & 0 & G & 0 & 0 \\ 0 & 0 & 0 & \kappa G & 0 \\ 0 & 0 & 0 & 0 & \kappa G \end{array} \right) \tag{4.73}
$$

$$[k_{ij}] = \frac{Eh}{1-\nu^2} \times$$

$$\int_A \begin{pmatrix} \frac{\partial N_i}{\partial x}\frac{\partial N_j}{\partial x} & \nu\frac{\partial N_i}{\partial x}\frac{\partial N_j}{\partial y} & 0 & 0 & 0 & 0 \\ \nu\frac{\partial N_i}{\partial y}\frac{\partial N_j}{\partial x} & \frac{\partial N_i}{\partial y}\frac{\partial N_j}{\partial y} & 0 & 0 & 0 & 0 \\ 0 & 0 & 0 & 0 & 0 & 0 \\ 0 & 0 & 0 & \frac{h^2}{12}\frac{\partial N_i}{\partial y}\frac{\partial N_j}{\partial y} & -\frac{\nu h^2}{12}\frac{\partial N_i}{\partial y}\frac{\partial N_j}{\partial x} & 0 \\ 0 & 0 & 0 & -\frac{\nu h^2}{12}\frac{\partial N_i}{\partial x}\frac{\partial N_j}{\partial y} & \frac{h^2}{12}\frac{\partial N_i}{\partial x}\frac{\partial N_j}{\partial x} & 0 \\ 0 & 0 & 0 & 0 & 0 & 0 \end{pmatrix} dA$$

$$+Gh \times$$

$$\int_A \begin{pmatrix} \frac{\partial N_i}{\partial y}\frac{\partial N_j}{\partial y} & \frac{\partial N_i}{\partial y}\frac{\partial N_j}{\partial x} & 0 & 0 \\ \frac{\partial N_i}{\partial x}\frac{\partial N_j}{\partial y} & \frac{\partial N_i}{\partial x}\frac{\partial N_j}{\partial x} & 0 & 0 \\ 0 & 0 & \kappa\left(\frac{\partial N_i}{\partial x}\frac{\partial N_j}{\partial x}+\frac{\partial N_i}{\partial y}\frac{\partial N_j}{\partial y}\right) & -\kappa\frac{\partial N_i}{\partial y}N_j \\ 0 & 0 & -\kappa N_i\frac{\partial N_j}{\partial y} & \frac{h^2}{12}\frac{\partial N_i}{\partial x}\frac{\partial N_j}{\partial x}+\kappa N_iN_j \\ 0 & 0 & \kappa N_i\frac{\partial N_j}{\partial x} & -\frac{h^2}{12}\frac{\partial N_i}{\partial y}\frac{\partial N_j}{\partial x} \\ 0 & 0 & 0 & 0 \end{pmatrix}$$

$$\begin{matrix} 0 & 0 \\ 0 & 0 \\ \kappa\frac{\partial N_i}{\partial x}N_j & 0 \\ -\frac{h^2}{12}\frac{\partial N_i}{\partial x}\frac{\partial N_j}{\partial y} & 0 \\ \frac{h^2}{12}\frac{\partial N_i}{\partial y}\frac{\partial N_j}{\partial y}+\kappa N_iN_j & 0 \\ 0 & 0 \end{matrix} \Bigg) dA$$

4.5.5.1 4-Node Rectangular Element

Consider a coordinate system in which the z-axis is located in the thickness direction of the plate and the x,y axes are located in the in-plane direction of the plate, as shown in **Fig. 4.17**. Also, considering two normal coordinates (ξ,η) oriented in the x,y direction in the plane of the plate, the shape function is the same as the shape function of a two-dimensional solid element as follows:

$$N_1 = \frac{1}{4}(1-\xi)(1-\eta), \qquad N_2 = \frac{1}{4}(1+\xi)(1-\eta),$$

$$N_3 = \frac{1}{4}(1-\xi)(1+\eta), \qquad N_4 = \frac{1}{4}(1+\xi)(1+\eta) \qquad (4.74)$$

4.5.5.2 Constant Strain Triangular Element

Considering a coordinate system in which the $z-$axis is located in the thickness direction of the plate and the x,y axes are located in the in-plane direction of the plate, as shown in **Fig. 4.18**. The shape function of the triangular element is as follows:

$$N_1 = \xi, \quad N_2 = \eta, \quad N_3 = \zeta \qquad (4.75)$$

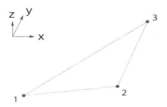

FIGURE 4.18
3-Node triangle element

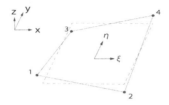

FIGURE 4.19
4-Node isoparametric quadrilateral element

4.5.5.3 Isoparametric Quadrilateral Element

Next, consider a 4-node quadrilateral element of arbitrary shape as shown in **Fig. 4.19**. In the case of a three-node element, the element shape is always flat, so the x, y axes can be placed in the plane of the plate, but if the four nodes are in free positions, the plane of the plate and shell element that passes through these nodal points will not be flat. Therefore, the x, y axes are taken in the plane tangent to the surface of the element at the center of the element, and the z-axis is taken in the direction orthogonal to this plane. Also, in general, the four nodes are located off this x-y plane in the z-axis direction, but we shall assume that this shift is small and treat the nodes as being at points projected onto the x-y plane. At this time, the expression of the isoparametric two-dimensional solid element can be used as is for the function representing the shape of the element, and the function representing the displacement.

4.5.6 Rigid Frame Element

Timoshenko shows the shape function of the beam element and the method of deriving the stiffness matrix based on the beam theory. Here, for later convenience, the strain and stress vectors are divided into the terms related to bending and shear deformation and the remaining terms, and are expressed as follows:

$$\varepsilon = \sum_{i=1}^{n} (B_i + B_i^s)^{\mathrm{T}} \boldsymbol{d}_i \ , \quad \sigma = \sum_{i=1}^{n} (DB_i + D^s B_i^s)^{\mathrm{T}} \boldsymbol{d}_i \qquad (4.76)$$

Here, B_i^s is the strain matrix due to bending-shear deformation, D^s is the elastic matrix for bending-shear deformation, B_i and D are the strain matrix and elastic matrix for the remaining terms.

From these, the stiffness matrix is expressed as follows:

$$[K_{ij}] = \int_V \left(B_i^{\mathrm{T}} D B_j + B_i^{s\,\mathrm{T}} D^s B_j^s \right) dV \qquad (4.77)$$

The nodal displacement vector, strain vector, stress vector, shape function vector, strain matrix, and elastic matrix are as follows:

$$\boldsymbol{d}_i = \left\{ \begin{array}{c} u_i \\ v_i \\ w_i \\ \hline \theta_{xi} \\ \theta_{yi} \\ \theta_{zi} \end{array} \right\}, \quad \boldsymbol{\varepsilon} = \left\{ \begin{array}{c} \varepsilon_{xx} \\ 2\varepsilon_{xy} \\ 2\varepsilon_{zx} \end{array} \right\}, \quad \boldsymbol{\sigma} = \left\{ \begin{array}{c} \sigma_{xx} \\ \sigma_{xy} \\ \sigma_{zx} \end{array} \right\} \qquad (4.78)$$

$$N_i = \left(\begin{array}{ccc|ccc} N_i & 0 & 0 & 0 & zN_i & -yN_i \\ 0 & N_i & 0 & -zN_i & 0 & 0 \\ 0 & 0 & N_i & yN_i & 0 & 0 \end{array} \right) \qquad (4.79)$$

$$B_i = \left(\begin{array}{ccc|ccc} \frac{dN_i}{dx} & 0 & 0 & 0 & z\frac{dN_i}{dx} & -y\frac{dN_i}{dx} \\ 0 & 0 & 0 & -z\frac{dN_i}{dx} & 0 & 0 \\ 0 & 0 & 0 & y\frac{dN_i}{dx} & 0 & 0 \end{array} \right),$$

$$B_i^s = \left(\begin{array}{ccc|ccc} 0 & 0 & 0 & 0 & 0 & 0 \\ 0 & \frac{dN_i}{dx} & 0 & 0 & 0 & -N_i \\ 0 & 0 & \frac{dN_i}{dx} & 0 & N_i & 0 \end{array} \right) \qquad (4.80)$$

$$D = \left(\begin{array}{ccc} E & 0 & 0 \\ 0 & G & 0 \\ 0 & 0 & G \end{array} \right), \quad D^s = \left(\begin{array}{ccc} 0 & 0 & 0 \\ 0 & \kappa_y G & 0 \\ 0 & 0 & \kappa_z G \end{array} \right) \qquad (4.81)$$

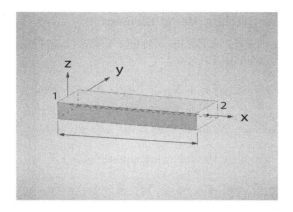

FIGURE 4.20
2-Node straight beam element

$$[K_{ij}] = \int_0^\ell \begin{pmatrix} EA\frac{dN_i}{dx}\frac{dN_j}{dx} & 0 & 0 & 0 & EG_y\frac{dN_i}{dx}\frac{dN_j}{dx} & -EG_z\frac{dN_i}{dx}\frac{dN_j}{dx} \\ 0 & 0 & 0 & 0 & 0 & 0 \\ 0 & 0 & 0 & 0 & 0 & 0 \\ 0 & 0 & 0 & GJ\frac{dN_i}{dx}\frac{dN_j}{dx} & 0 & 0 \\ EG_y\frac{dN_i}{dx}\frac{dN_j}{dx} & 0 & 0 & 0 & EI_y\frac{dN_i}{dx}\frac{dN_j}{dx} & -EI_{yz}\frac{dN_i}{dx}\frac{dN_j}{dx} \\ -EG_z\frac{dN_i}{dx}\frac{dN_j}{dx} & 0 & 0 & 0 & -EI_{yz}\frac{dN_i}{dx}\frac{dN_j}{dx} & EI_z\frac{dN_i}{dx}\frac{dN_j}{dx} \end{pmatrix} dx$$

$$+GA\int_0^\ell \begin{pmatrix} 0 & 0 & 0 & 0 & 0 & 0 \\ 0 & \kappa_y\frac{dN_i}{dx}\frac{dN_j}{dx} & 0 & 0 & 0 & -\kappa_y\frac{dN_i}{dx}N_j \\ 0 & 0 & \kappa_z\frac{dN_i}{dx}\frac{dN_j}{dx} & 0 & \kappa_z\frac{dN_i}{dx}N_j & 0 \\ 0 & 0 & 0 & 0 & 0 & 0 \\ 0 & 0 & \kappa_z N_i\frac{dN_j}{dx} & 0 & \kappa_z N_i N_j & 0 \\ 0 & -\kappa_y N_i\frac{dN_j}{dx} & 0 & 0 & 0 & \kappa_y N_i N_j \end{pmatrix} dx$$

where,

$$A = \int_A dA , \quad G_y = \int_A z\,dA , \quad G_z = \int_A y\,dA \qquad (4.82)$$

$$I_y = \int_A z^2 dA , \quad I_z = \int_A y^2 dA , \quad I_{yz} = \int_A yz\,dA , \qquad (4.83)$$

$$J = \int_A (y^2 + z^2)dA \qquad (4.84)$$

4.5.6.1   2-Node Straight Beam Element

The displacement of the two-node straight beam element as shown in **Fig. 4.20** can be expressed as follows:

$$
\begin{aligned}
u &= N_1 u_1 + N_2 u_2, & \theta_x &= N_1 \theta_{x1} + N_2 \theta_{x2} \\
v &= N_1 v_1 + N_2 v_2, & \theta_y &= N_1 \theta_{y1} + N_2 \theta_{y2} \\
w &= N_1 w_1 + N_2 w_2, & \theta_z &= N_1 \theta_{z1} + N_2 \theta_{z2}
\end{aligned}
$$

Here, N_1, N_2 are given by the following equations.

$$
N_1 = 1 - \xi, \quad N_2 = \xi, \quad \xi = x/\ell \tag{4.85}
$$

The stiffness matrix obtained from this displacement function is as follows:

$$
K_{11} = \begin{pmatrix}
EA/\ell & & & & EG_y/\ell & -EG_z/\ell \\
& \kappa_y GA/\ell & & & & \kappa_y GA/2 \\
& & \kappa_z GA/\ell & -\kappa_z GA/2 & & \\
& & & GJ/\ell & & \\
EG_y/\ell & & -\kappa_z GA/2 & & EI_y/\ell + \kappa_z GA\ell/3 & -EI_{yz}/\ell \\
-EG_z/\ell & \kappa_y GA/2 & & & -EI_{yz}/\ell & EI_z/\ell + \kappa_y GA/3
\end{pmatrix}
$$

$$
K_{22} = \begin{pmatrix}
EA/\ell & & & & EG_y/\ell & -EG_z/\ell \\
& \kappa_y GA/\ell & & & & -\kappa_y GA/2 \\
& & \kappa_z GA/\ell & \kappa_z GA/2 & & \\
& & & GJ/\ell & & \\
EG_y/\ell & & \kappa_z GA/2 & & EI_y/\ell + \kappa_z GA\ell/3 & -EI_{yz}/\ell \\
-EG_z/\ell & -\kappa_y GA/2 & & & -EI_{yz}/\ell & EI_z/\ell + \kappa_y GA/3
\end{pmatrix}
$$

$$
K_{12} = \begin{pmatrix}
-EA/\ell & & & & -EG_y/\ell & EG_z/\ell \\
& -\kappa_y GA/\ell & & & & \kappa_y GA/2 \\
& & -\kappa_z GA/\ell & -\kappa_z GA/2 & & \\
& & & -GJ/\ell & & \\
-EG_y/\ell & \kappa_z GA/2 & & & -EI_y/\ell + \kappa_z GA\ell/6 & EI_{yz}/\ell \\
EG_z/\ell & -\kappa_y GA/2 & & & EI_{yz}/\ell & -EI_z/\ell + \kappa_y GA\ell/6
\end{pmatrix}
$$

However, when the beam member is solved using this stiffness matrix, the accuracy is inferior except for the beam with a small slenderness ratio, and it cannot be applied to the analysis of the actual member. This is due to a decrease in accuracy called **shear locking**, and for 2-node elements, the accuracy is improved when the integral of the stiffness matrix is expressed as the product of the integral value at the center of the element and the element length, as follows. Such an integration method is called the reduced integration method.

$$
\int_0^\ell F(x)dx \approx F(\ell/2)\ell \tag{4.86}
$$

When this reduced integration is performed, the coefficients of 1/3, 1/6 are replaced with 1/4 in the above stiffness matrix.

4.6   Finite Element Method Techniques

4.6.1   Numerical Integration Method

The integral of the element stiffness matrix becomes more complicated with higher order elements. In addition, if the element shape is distorted in the element using the parametric map, the calculation of the integral becomes more difficult. In such cases, an approximate calculation of the integral is required.

When calculating the integral approximately, the integral $f(\xi)$ is calculated with the function value $f(\xi_i)$ calculated at some points on the domain $(-1 \leq \xi \leq 1)$ as shown in the following equation. It is expressed by multiplying it by an appropriate weighting factor W_i and adding it.

$$\int_{-1}^{1} f(\xi) d\xi \approx \sum_{i=1}^{n} w_i f(\xi_i) \tag{4.87}$$

Various integration methods depends on how to select the integration points $\xi_1, \xi_2, \cdots, \xi_n$ and the weighting coefficients $w_1, w_2, \cdots, w_n$ in the above equation. Here, two typical integration methods are described.

4.6.1.1   Newton-Cotes Numerical Integration Method

Select integration points $\xi_1, \xi_2, \cdots, \xi_n$ at equal intervals, with the $(n-1)$ next multinormal $f_n(\xi_i)$ that matches the function $f(\xi_i)$ in these points expressed by the Lagrange interpolation formula as follows:

$$f_n(\xi) \equiv \sum_{i=1}^{n} N_i(\xi) f(\xi_i) \tag{4.88}$$

Instead of integrating the function $f(\xi_i)$,

$$\int_{-1}^{1} f(\xi) \, d\xi \simeq \int_{-1}^{1} f_a(\xi) \, d\xi$$

considering the integral, the weighting coefficient of Eq.(5.1) is obtained as follows:

$$w_i = \int_{-1}^{1} N_i(\xi) d\xi = \int_{-1}^{1} \prod_{j=1}^{n} \frac{\xi - \xi_j}{\xi_i - \xi_j} d\xi \tag{4.89}$$

This integration method is called trapezoidal integration when the integration point n is 2 and Simpson's rule when the integration point n is 3.

4.6.1.2   Gauss-Legendre Numerical Integration Method

This method does not fix the integration points at equally spaced points but determines the coordinate values and weighting factors of the integration points so as to give an accurate integration value. In this case, $f_a(\xi_i)$ is expressed as a function determined by a total of $2n$ of unknowns, which is the sum of the coordinate value ξ_i and the weighting coefficient w_i at the integration points of n. Therefore, $f_a(\xi_i)$ is represented as a polynomial of order $(2n-1)$.

Now, a Lagrange interpolation function matches the integrand $f_a(\xi_i)$ at n points to the integrand $f(\xi_i)$. If becoming zero at these points, the high-order polynomials are expressed as follows:

$$f_a(\xi) = \sum_{i=1}^{n} N_i(\xi) f(\xi_i) + N_0(\xi)(\alpha_0 + \alpha_1 \xi + \cdots + \alpha_{n-1} \xi^{n-1}) \qquad (4.90)$$

Here, $N_0(\xi)$ is a function that becomes zero at n integration points.

Integrating this equation instead of $f(\xi_i)$

$$\int_{-1}^{1} f(\xi) d\xi \approx \int_{-1}^{1} f_a(\xi) d\xi = \sum_{i=1}^{n} \int_{-1}^{1} N_i(\xi) d\xi f(\xi_i) + \sum_{i=0}^{n-1} \alpha_i \int_{-1}^{1} N_0(\xi) \xi^i d\xi$$
$$(4.91)$$

use the Legendre polynomial $P_n(\xi)^2$ of the following n for $N_0(\xi)$ in the above equation. From orthogonality,

$$\int_{-1}^{1} P_n(\xi) \xi^m d\xi = 0 \quad (m = 0, 1, \ldots, n-1) \qquad (4.92)$$

The second term on the right-hand side of Eq.(4.91) disappears, and the weighting coefficient w_i is obtained by comparison with Eq.(4.87), which is obtained as an equation similar to Eq.(4.89). However, it is necessary to use the point where the Legendre polynomial $P_n(\xi)$ becomes zero as the integration point. The Newton-Cotes integration method described above can accurately integrate up to $(n-1)$ order polynomials at n integration points. Furthermore, this Gauss-Legendre integral method can accurately integrate up to $(2n-1)$ order polynomials, and this method is often used for the integration of stiffness matrices, and so on.

4.6.1.3   Numerical Integration in Multiple Dimensions

Even when calculating multiple integrals in a two-dimensional or three-dimensional domain numerically, the position of the integration point and the weighting coefficient can be determined so that the integral is the most accurate. The simplest and most commonly used method is to apply the one-dimensional Newton-Cotes integral method or Gauss-Legendre integral method for each variable.

However, in the integration of the triangular and tetrahedral regions, if the combination of the one-dimensional integration methods is used as is, because of the geometrical directionality, it is better to use the position of the integration point and the weighting coefficient to obtain an accurate integration in these regions.

4.6.2   Mesh Dependency of Solution

The greatest feature of the Finite Element Method is that it can analyze objects with complex boundary shapes. Recently, various automatic element division methods that automatically perform finite element division have been devised, and pre-processors incorporating this function have become widespread. Therefore, it is possible to divide the mesh without much effort.

However, it should be noted that the solution by the finite element method changes depending on how to divide the mesh. Of course, if the element division is made more refined, the influence of the mesh division will be lessened, but unlimited division is not possible due to various restrictions.

If the structure is symmetric and the load is also symmetric, then the displacement and stress should be symmetric or inversely symmetric with respect to this axis of symmetry, but if the mesh division is not symmetric, this is not the case.

Also, if there is a cut in the boundary or a hole in the interior, the mesh division needs to be finer than in other parts of the solution because the solution changes rapidly in the vicinity of the cut, but in areas where the solution does not change so much, the mesh division does not need to be so fine. In other words, the finite element method can be used to solve any complicated problem, but in order to obtain a reliable solution, the mechanical sensibility of the engineer is an essential requirement.

Therefore, first, starting from an appropriate mesh division, an error estimation of each element is performed, and this estimation error is reflected in the mesh division. Then, the mesh division, which is called the Adaptive Remeshing Method, is repeatedly corrected to improve the reliability of the solution.

4.7   Programming the Finite Element Method

4.7.1   Assembly of Overall Stiffness Matrix

Until now, we have explained how to derive the element stiffness equations regarding various structural members. And because the actual structure is a complicated structure in which beams, columns, and plates are joined, it is necessary to join the individual element stiffness equations to create the overall stiffness equation. Moreover, since the finite element method is an approximate solution method, an accurate solution can not be obtained even in a single member, in which multiple elements are used for element division. Therefore, here we explain how to derive the stiffness equation for the overall structure from the element stiffness equation.

Consider the structure of a 3-node structure consisting of two elements as shown in **Fig. 4.23**. The stiffness equations for the individual elements are as

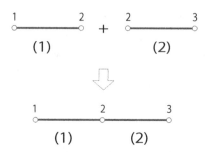

FIGURE 4.21
4 deformation mode of node element

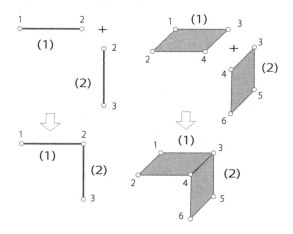

FIGURE 4.22
Joining members with different element orientations

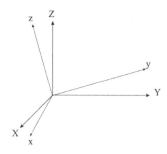

FIGURE 4.23
Coordinate transformation

follows:

$$
\left\{\begin{array}{c} P_1^{(1)} \\ P_2^{(1)} \end{array}\right\} = \left(\begin{array}{cc} k_{11}^{(1)} & k_{12}^{(1)} \\ k_{21}^{(1)} & k_{22}^{(1)} \end{array}\right)\left\{\begin{array}{c} d_1 \\ d_2 \end{array}\right\}, \quad \left\{\begin{array}{c} P_2^{(2)} \\ P_3^{(2)} \end{array}\right\} = \left(\begin{array}{cc} k_{22}^{(2)} & k_{23}^{(2)} \\ k_{32}^{(2)} & k_{33}^{(2)} \end{array}\right)\left\{\begin{array}{c} d_2 \\ d_3 \end{array}\right\}
$$

$$(4.93)$$

Here, d_1 and d_2 are nodes, respectively. There is a displacement vector of 1, 2 and 3, and the displacement of node 2 of element (1) and the displacement of node 2 of element (2) are equal because the elements are connected (conformity condition).

In addition, $P_1^{(1)}$ and $P_2^{(1)}$ are the load vectors of node 1 and node 2 of the element (1), and $P_2^{(2)}$ and $P_3^{(3)}$ are the load vectors of node 2 and node 3 of the element (2).

Now, when the load vector of each node and P_1, P_2, and P_3 are expressed, the following relationship must be established from the equilibrium equation at each node;

$$
P_1 = P_1^{(1)}, \quad P_2 = P_2^{(1)} + P_2^{(2)}, \quad P_3 = P_3^{(2)} \tag{4.94}
$$

Substituting the stiffness equation for each element into the equation gives the following equation;

$$
\left\{\begin{array}{c} P_1 \\ P_2 \\ P_3 \end{array}\right\} = \left(\begin{array}{ccc} k_{11}^{(1)} & k_{12}^{(1)} & 0 \\ k_{21}^{(1)} & k_{22}^{(1)}+k_{22}^{(2)} & k_{23}^{(2)} \\ 0 & k_{32}^{(2)} & k_{33}^{(2)} \end{array}\right)\left\{\begin{array}{c} d_1 \\ d_2 \\ d_3 \end{array}\right\} \tag{4.95}
$$

This is the stiffness equation for the entire structure in **Fig. 4.21**. In this way, by using the displacement matching conditions and equilibrium conditions at each node, the stiffness equation of the entire structure can be obtained from the individual element stiffness equations.

4.7.2 Coordinate Transformation

If the coordinate system for deriving the element stiffness equation is different for each element, the stiffness equation of the entire structure can not be derived unless the element stiffness equation is coordinate transformed in the coordinate system of the entire structure and the directions of the components of the displacement vector and the load vector are matched. In the case of solid elements, it may be necessary to set individual element coordinate systems to perform operations to mitigate locking.

In addition, for beam elements and plate and shell elements, in order to facilitate guidance, since individual element coordinate systems are set and the element stiffness equations are derived in this coordinate system, the coordinates must be transformed into the coordinate system of the entire structure. Now, assume that the element coordinate system is (x, y, z) and the overall coordinate system is (X, Y, Z). The relationship between the vector $\boldsymbol{x}$ is represented by the components of the (x, y, z) coordinate system and vector $\boldsymbol{X}$ by the components of the (X, Y, Z) coordinate system. It is expressed as follows:

$$\boldsymbol{x} = T\boldsymbol{X} \tag{4.96}$$

Here, T is a matrix that represents the relationship between two coordinate systems and is called a coordinate transformation matrix;

$$T^{-1} = T^{\mathrm{T}} \tag{4.97}$$

If the coordinate system is orthogonal, the following relationship holds for the inverse matrix of this matrix.

Therefore, the inverse relationship of Eq.(4.96) is expressed as follows:

$$\boldsymbol{X} = T^{\mathrm{T}}\boldsymbol{x} \tag{4.98}$$

Considering these relationships,

$$k\boldsymbol{d} = \boldsymbol{p}$$

consider expressing the stiffness equation expressed in the element coordinate system in the overall coordinate system. Now, if the nodal displacement vector $\boldsymbol{d}$, the load vector $\boldsymbol{p}$, and the stiffness matrix k are represented by the components of the global coordinate system as $\boldsymbol{D}, \boldsymbol{P}$, and K, respectively, in the global coordinate system, the stiffness equation is

$$K\boldsymbol{D} = \boldsymbol{P}$$

At this time, the relationship between these two equations is

$$\boldsymbol{d} = T\boldsymbol{D} , \quad \boldsymbol{P} = T^{\mathrm{T}}\boldsymbol{p} \tag{4.99}$$

because

$$\boldsymbol{P} = T^{\mathrm{T}}\boldsymbol{p} = T^{\mathrm{T}}k\boldsymbol{d} = T^{\mathrm{T}}kT\boldsymbol{D}. \tag{4.100}$$

Therefore, the stiffness matrix K in the global coordinate system is expressed as follows from the stiffness matrix in the element coordinate system;

$$K = T^{\mathrm{T}}kT. \tag{4.101}$$

4.7.3   Introduction Method of Support Conditions

The stiffness equation derived in the previous section does not give boundary conditions for displacement such as support conditions. Therefore, the boundary conditions for displacement will be explained. It is assumed that the stiffness equation of the entire structure is expressed as follows:

$$
\begin{pmatrix}
k_{11} & k_{12} & \cdots & k_{1m} & \cdots & k_{1n} \\
k_{21} & k_{21} & \cdots & k_{2m} & \cdots & k_{2n} \\
\vdots & \vdots & \ddots & \vdots & & \vdots \\
k_{m1} & k_{m2} & \cdots & k_{mm} & \cdots & k_{mn} \\
\vdots & \vdots & & \vdots & \ddots & \vdots \\
k_{n1} & k_{n2} & \cdots & k_{nm} & \cdots & k_{nn}
\end{pmatrix}
\begin{Bmatrix}
d_1 \\ d_2 \\ \vdots \\ d_m \\ \vdots \\ d_n
\end{Bmatrix}
=
\begin{Bmatrix}
p_1 \\ p_2 \\ \vdots \\ p_m \\ \vdots \\ p_n
\end{Bmatrix}
\tag{4.102}
$$

Now, consider the case where the displacement d_m of the m component is constrained and $d_m = 0$. At this time, a reaction force r_m in the d_m direction is generated, so the stiffness equation in the above equation is as follows:

$$
\begin{pmatrix}
k_{11} & k_{12} & \cdots & k_{1m} & \cdots & k_{1n} \\
k_{21} & k_{21} & \cdots & k_{2m} & \cdots & k_{2n} \\
\vdots & \vdots & \ddots & \vdots & & \vdots \\
k_{m1} & k_{m2} & \cdots & k_{mm} & \cdots & k_{mn} \\
\vdots & \vdots & & \vdots & \ddots & \vdots \\
k_{n1} & k_{n2} & \cdots & k_{nm} & \cdots & k_{nn}
\end{pmatrix}
\begin{Bmatrix}
d_1 \\ d_2 \\ \vdots \\ 0 \\ \vdots \\ d_n
\end{Bmatrix}
=
\begin{Bmatrix}
p_1 \\ p_2 \\ \vdots \\ r_m \\ \vdots \\ p_n
\end{Bmatrix}
\tag{4.103}
$$

$d_i (i = 1, 2, \cdots, m - 1, m + 1, \cdots, n)$ and r_m are unknown quantities. $p_i (i = 1, 2, \cdots, m - 1, m + 1, \cdots, n)$ and $d_m (= 0)$ are known quantities. Because unknown and known quantities are mixed on both sides in this way, programming becomes complicated, the two above equations are divided into the following;

$$
\begin{pmatrix}
k_{11} & k_{12} & \cdots & 0 & \cdots & k_{1n} \\
k_{21} & k_{21} & \cdots & 0 & \cdots & k_{2n} \\
\vdots & \vdots & \ddots & \vdots & & \vdots \\
0 & 0 & \cdots & 1 & \cdots & 0 \\
\vdots & \vdots & & \vdots & \ddots & \vdots \\
k_{n1} & k_{n2} & \cdots & 0 & \cdots & k_{nn}
\end{pmatrix}
\begin{Bmatrix}
d_1 \\ d_2 \\ \vdots \\ d_m \\ \vdots \\ d_n
\end{Bmatrix}
=
\begin{Bmatrix}
p_1 \\ p_2 \\ \vdots \\ 0 \\ \vdots \\ p_n
\end{Bmatrix}
\tag{4.104}
$$

$$
\begin{pmatrix} k_{m1} & k_{m2} & \cdots & k_{mm} & \cdots & k_{mn} \end{pmatrix}
\begin{Bmatrix}
d_1 \\ d_2 \\ \vdots \\ d_m \\ \vdots \\ d_n
\end{Bmatrix}
= r_m
\tag{4.105}
$$

If solving the first equation of the above equation, a solution of the node displacement considering the boundary conditions will be obtained, and then, substituting the solution into the second equation, the reaction force can be obtained. In the case of multiple constraints, the procedure for deriving the above equation may be better by using multiple constraints conditions.

4.7.4   Solving Simultaneous Equations

The numerical calculation method for finding the vector x from the following simultaneous equations is described;

$$Ax = b \qquad (4.106)$$

Numerical solutions to simultaneous equations can generally be divided into elimination (direct method) and the iterative method.

- countermeasure method
  A solution method that starts from a certain initial value and improves the accuracy of the solution by iterative calculation. Due to the characteristics of the coefficient matrix A, it has the disadvantage that the number of operations to obtain a solution with the required accuracy differs, but it is suitable for solving large-dimensional problems.

- elimination method (direct method)
  A method that can obtain a solution with a finite number of operations, and the calculation time can be predicted. However, it is not suitable for solving large-dimensional problems.

4.7.4.1   Iteration

Among iteration methods, for thinking and programming easily, the simultaneous iterative method (Jacobi method) and the successive iterative method (Gauss-Seidel method) are explained. Now, the initial value of the solution is expressed as $x^{(0)}$. This can be of any value. The number of operations changes depending on this initial value, but all components may be set to zero, for example, $x^{(0)}=0$. The simultaneous iterative method (Jacobi method) is a method in which the coefficient matrix A is divided into a matrix A_d in which only diagonal elements are extracted and a matrix A_r consisting of the remaining elements, and the following equation is repeatedly solved;

$$A_d x^{(k)} = -A_r x^{(k-1)} + b, \quad k>1 \qquad (4.107)$$

Here,

$$A_d = \begin{pmatrix} a_{11} & 0 & 0 & \cdots & 0 \\ 0 & a_{22} & 0 & \cdots & 0 \\ 0 & 0 & a_{33} & \cdots & 0 \\ \vdots & \vdots & \vdots & \ddots & \vdots \\ 0 & 0 & 0 & \cdots & a_{nn} \end{pmatrix}, \quad A_T = \begin{pmatrix} 0 & a_{12} & a_{13} & \cdots & a_{1n} \\ a_{21} & 0 & a_{23} & \cdots & a_{2n} \\ a_{31} & a_{32} & 0 & \cdots & a_{3n} \\ \vdots & \vdots & \vdots & \ddots & \vdots \\ a_{n1} & a_{n2} & a_{n3} & \cdots & 0 \end{pmatrix}$$
$$(4.108)$$

The successive iterative method (Gauss-Seidel method) is divided into the matrix A_u, which extracts only the upper-right element from the diagonal element, and the matrix A_l, which extracts only the lower-left element from the above matrix A. This is a method to solve the following equation repeatedly;

$$A_d x^{(k)} = -A_l x^{(k)} - A_u x^{(k-1)} + b, \quad k > 1 \tag{4.109}$$

Here,

$$A_u = \begin{pmatrix} 0 & a_{12} & a_{13} & \cdots & a_{1n} \\ 0 & 0 & a_{23} & \cdots & a_{2n} \\ 0 & 0 & 0 & \cdots & a_{3n} \\ \vdots & \vdots & \vdots & \ddots & \vdots \\ 0 & 0 & 0 & \cdots & 0 \end{pmatrix}, \quad A_l = \begin{pmatrix} 0 & a_{12} & a_{13} & \cdots & a_{1n} \\ a_{21} & 0 & a_{23} & \cdots & a_{2n} \\ a_{31} & a_{32} & 0 & \cdots & a_{3n} \\ \vdots & \vdots & \vdots & \ddots & \vdots \\ a_{n1} & a_{n2} & a_{n3} & \cdots & 0 \end{pmatrix} \tag{4.110}$$

When finding the i component $x_i^{(k)}$ of the vector $x^{(k)}$ in the simultaneous iterative method, this method uses the new approximation $x_1^{(k)}, \cdots, x_{i-1}^{(k)}$ that has already been obtained, and converges faster than the simultaneous iterative method.

4.7.4.2 Elimination Method (Direct Method)

There are various elimination methods, but the LU decomposition method, which is often used in the Finite Element Method, will be explained here. In the LU decomposition method, the coefficient matrix A is represented by the product of the lower triangular matrix L, the diagonal matrix D, and the upper triangular matrix U as follows:

$$A = LDU \tag{4.111}$$

here,

$$L = \begin{pmatrix} 1 & 0 & 0 & \cdots & 0 \\ l_{21} & 1 & 0 & \cdots & 0 \\ l_{31} & l_{32} & & \cdots & 0 \\ \vdots & \vdots & \vdots & \ddots & \vdots \\ l_{n1} & l_{n2} & l_{n3} & \cdots & 1 \end{pmatrix}, \quad D = \begin{pmatrix} d_1 & 0 & 0 & \cdots & 0 \\ 0 & d_2 & 0 & \cdots & 0 \\ 0 & 0 & d_3 & \cdots & 0 \\ \vdots & \vdots & \vdots & \ddots & \vdots \\ 0 & 0 & 0 & \cdots & d_n \end{pmatrix},$$

$$U = \begin{pmatrix} 1 & u_{12} & u_{13} & \cdots & u_{1n} \\ 0 & 1 & u_{23} & \cdots & u_{2n} \\ 0 & 0 & 1 & \cdots & u_{3n} \\ \vdots & \vdots & \vdots & \ddots & \vdots \\ 0 & 0 & 0 & \cdots & 1 \end{pmatrix} \tag{4.112}$$

$$Ly = b, \quad DUx = y \tag{4.113}$$

These equations can be easily solved, the components of the vectors y, x and b are expressed as y_i, x_i, b_i $(i = 1, 2, \cdots, n)$, and it is as follows:

$$y_1 = b_1, \quad y_i = b_i - \sum_{k=1}^{i-1} l_{ik} y_k, \quad i = 2, 3, \cdots, n \tag{4.114}$$

$$x_n = \frac{y_n}{d_n}, \quad x_i = \frac{y_i}{d_i} - \sum_{k=i+1}^{n} u_{ik}x_k, \quad i = n-1, \cdots, 2, 1 \qquad (4.115)$$

It should be noted that, in most of the simultaneous equations handled by the finite element method, the coefficient matrix is a symmetric matrix. Then L disassembled by the LU is equal to U^{T}. Therefore, in a symmetric matrix,

$$A = LDL^{\mathrm{T}} = U^{\mathrm{T}}DU \qquad (4.116)$$

This is expressed as an element as follows:

$$a_{ij} - \sum_{k=1}^{i} u_{ki}d_k u_{kj}, \quad i \leq j \qquad (4.117)$$

Note that $u_{ii} = 1$. From this, the following relational expression can be obtained.

$$d_1 = a_{11}, \quad u_{1j} = \frac{a_{1j}}{d_2}, \quad j = 2, 3, \cdots, n \qquad (4.118)$$

$$d_i = a_{ii} - \sum_{k=1}^{i-1} u_{ki}d_k u_{ki},$$

$$u_{ij} = \frac{1}{d_1}\left(a_{ij} - \sum_{k=1}^{i-1} u_{ki}d_k u_{kj}\right), \quad j = i+1, i+2, \cdots, n \quad (4.119)$$

Part II

This is the Application of Group Theory to Symmetric Structures

5

Block Diagonalization Theory for Dihedral Groups

In this chapter, we introduce block diagonalization methods such as stiffness matrix, damping matrix, and mass matrix that transform the geometric properties of finite element models of symmetric structures with the dihedral group D_n of finite regular n-gonal symmetry.

5.1 Dihedral Group D_n and Irreducible Representation

Block diagonalization theory is a general theory that holds for various groups, however, here we restrict our discussion to dihedral groups.

The dihedral group D_n, which represents the regular n-gonometric symmetry, is generated from the rotational transformation r and the mirroring transformation s and is defined as

$$D_n \equiv \{r^k, sr^k \mid k = 0, \cdots, n-1\}, \qquad (5.1)$$
$$C_n \equiv \{r^k \mid k = 0, \cdots, n-1\} \qquad (5.2)$$

where $r^n = s^2 = (sr)^2 = 1$ and where 1 denotes the identity transformation s represents the mirror transformation about the X-axis, and r^j is the counterclockwise transformation around the Z-axis $2j\pi/n$ $(j = 1, \cdots, n-1)$ rotational transformations, respectively.

With this transformation r^k, the counterclockwise rotation operation around a certain origin can be expressed as

$$R = \begin{pmatrix} \cos k\theta & -\sin k\theta & 0 \\ \sin k\theta & \cos k\theta & 0 \\ 0 & 0 & 1 \end{pmatrix}, \quad \theta = 2\pi/n \qquad (5.3)$$

and matrix, and any point $\boldsymbol{x} = (x, y, z)^{\mathrm{T}}$ in 3D space is transformed as

$$r : \boldsymbol{x} \longrightarrow R\,\boldsymbol{x} \qquad (5.4)$$

by this rotation operation. Also, the mirroring operation on the XZ plane by

DOI: 10.1201/9781032670386-5

the transformation s is represented by the matrix

$$S = \begin{pmatrix} 1 & 0 & 0 \\ 0 & -1 & 0 \\ 0 & 0 & 1 \end{pmatrix} \tag{5.5}$$

and the point $\boldsymbol{x}$ is transformed to

$$s : \boldsymbol{x} \longrightarrow S\boldsymbol{x} \tag{5.6}$$

by this mirroring operation.

The D_n-invariant subgroup is defined to be

$$D_m^j \equiv \{r^{kn/m}, sr^{kn/m+j-1} \mid k = 0, \cdots, m-1\} \tag{5.7}$$
$$C_m \equiv \{r^{kn/m} \mid k = 0, \cdots, m-1\} \tag{5.8}$$

where $D_m = D_m^1$ and $C_1 = \{1\}$, and let $m = \gcd(j, n)$ denote the greatest common divisor of j and n. The Dihedral group D_m^j has linear symmetry with m axes. The cyclic group C_m represents rotational symmetry with regard to the angle $2\pi/m$.

We denote by

$$\mathcal{R}(D_n) = \{\mu \equiv (d, j) \mid j = 1, \cdots, m_d; \quad d = 1, 2\} \tag{5.9}$$

the entire set of irreducible representations of the Dihedral group D_n (for details on symbols, see Murota and Ikeda [27, 28]). In $\mu = (d, j)$, d denotes the order of the irreducible representation μ, and j denotes the j-th d-order irreducible representation. Also, m_d is the number of d-order irreducible representations.

$$\begin{cases} m_1 = 4, & m_2 = n/2 - 1, & \text{when } n = \text{even} \\ m_1 = 2, & m_2 = (n-1)/2, & \text{when } n = \text{odd} \end{cases} \tag{5.10}$$

The first-order irreducible representation matrix of D_n for the rotational transformation r and the mirroring transformation s is uniquely determined to be

$$\begin{array}{llll} T^{(1,1)}(r) = & 1, & T^{(1,1)}(s) = & 1; \\ T^{(1,2)}(r) = & 1, & T^{(1,2)}(s) = & -1; \\ T^{(1,3)}(r) = & -1, & T^{(1,3)}(s) = & 1; \\ T^{(1,4)}(r) = & -1, & T^{(1,4)}(s) = & -1. \end{array} \tag{5.11}$$

Although the second-order irreducible representation matrix is not uniquely determined, it is defined as

$$T^{(2,j)}(r) = \begin{pmatrix} \cos(2\pi j/n) & -\sin(2\pi j/n) \\ \sin(2\pi j/n) & \cos(2\pi j/n) \end{pmatrix}, \tag{5.12}$$

$$T^{(2,j)}(s) = \begin{pmatrix} 1 & 0 \\ 0 & -1 \end{pmatrix}. \tag{5.13}$$

according to Murota and Ikeda [27, 28]. In this case, the indicator is

$$\chi^{(2,j)}(r) = 2\cos(2\pi j/n), \tag{5.14}$$
$$\chi^{(2,j)}(s) = 0 \tag{5.15}$$

5.2 Block Diagonalization

5.2.1 Static Equilibrium Equation

Define

$$
\begin{aligned}
H &\equiv [\cdots, H^{\mu}, \cdots] \\
&= [\; H^{(1,1)}, \;\; \cdots, H^{(1,m_1)}, \\
&\qquad H^{(2,1)+}, \;\; \cdots, H^{(2,m_2)+}, \\
&\qquad H^{(2,1)-}, \;\; \cdots, H^{(2,m_2)-} \;]
\end{aligned}
\tag{5.16}
$$

as the coordinate transformation matrix that decomposes the equilibrium equation (3.17) for each irreducible representation. where $H^{(1,j)}$ denotes the partial block matrix corresponding to the first order vested representation $(1, j)$, and $H^{(2,j)+}$ and $H^{(2,j)-}$ denote the partial block matrices corresponding to the second-order irreducible representation $(2, j)$, respectively. Using this coordinate transformation matrix H, we obtain the stiffness matrix which can be block-diagonalized to

$$
\begin{aligned}
\widetilde{K} &= H^{\mathrm{T}} K H = \mathrm{diag}[\cdots, \widetilde{K}^{\mu}, \cdots] \\
&= \mathrm{diag}[\widetilde{K}^{(1,1)}, \;\; \cdots, \widetilde{K}^{(1,m_1)}, \\
&\qquad \widetilde{K}^{(2,1)+}, \;\; \cdots, \widetilde{K}^{(2,m_2)+}, \\
&\qquad \widetilde{K}^{(2,1)-}, \;\; \cdots, \widetilde{K}^{(2,m_2)-} \;]
\end{aligned}
\tag{5.17}
$$

where $\mathrm{diag}[\cdots]$ denotes the block diagonal matrix. However, it is

$$
\widetilde{K}^{(2,j)+} = \widetilde{K}^{(2,j)-}, \quad j = 1, \cdots, m_2
\tag{5.18}
$$

Two identical diagonal blocks correspond to a quadratic irreducible representation. Due to the orthogonality between the irreducible representations, each block matrix is given by

$$
\widetilde{K}^{\mu} = (H^{\mu})^{\mathrm{T}} K H^{\mu}, \quad \mu \in \mathcal{R}(D_n)
\tag{5.19}
$$

which is used for the actual calculation.

The coordinate system corresponding to the irreducible representation is defined as

$$
\begin{aligned}
\boldsymbol{u} &= H\boldsymbol{w} = \sum_{\mu \in \mathcal{R}(G)} H^{\mu} \boldsymbol{w}^{\mu} \\
&= \sum_{j=1}^{m_1} H^{(1,j)} \boldsymbol{w}^{(1,j)} + \sum_{j=1}^{m_2} (H^{(2,j)+} \boldsymbol{w}^{(2,j)+} + H^{(2,j)-} \boldsymbol{w}^{(2,j)-})
\end{aligned}
\tag{5.20}
$$

Here, the variable

$$
\begin{aligned}
\boldsymbol{w} = [\; &(\boldsymbol{w}^{(1,1)})^{\mathrm{T}}, \;\; \cdots, (\boldsymbol{w}^{(1,m_1)})^{\mathrm{T}}, \\
&(\boldsymbol{w}^{(2,1)+})^{\mathrm{T}}, \cdots, (\boldsymbol{w}^{(2,m_2)+})^{\mathrm{T}}, \\
&(\boldsymbol{w}^{(2,1)-})^{\mathrm{T}}, \cdots, (\boldsymbol{w}^{(2,m_2)-})^{\mathrm{T}} \;]^{\mathrm{T}}
\end{aligned}
\tag{5.21}
$$

in the new coordinate system can be expressed as a variable for each irreducible representation. Using Eq.(5.20) to transform the coordinates, the equilibrium equation (3.17) can be decomposed into a form corresponding to the following;

$$(H^{(1,j)})^{\mathrm{T}}f = \tilde{K}^{(1,j)}w^{(1,j)},$$
$$(H^{(2,j)+})^{\mathrm{T}}f = \tilde{K}^{(2,j)+}w^{(2,j)+},$$
$$(H^{(2,j)-})^{\mathrm{T}}f = \tilde{K}^{(2,j)-}w^{(2,j)-},$$
$$j = 1,\cdots,m_d \; ; \quad d = 1,2 \tag{5.22}$$

and the irreducible representation equation (5.20). Substituting the solution of Eq.(5.22) into Eq.(5.20), we obtain the solution u as a superposition of the solutions of each block.

The symmetry of the column vector of the matrix H^μ is defined as

$$\Sigma(H^\mu) = \{g \in D_n \mid T^\mu(g) = \mathrm{I}\}. \tag{5.23}$$

Here, $T^\mu(g)$ is the representation matrix for the irreducible representation μ. The relation between the coordinate transformation matrix and the symmetry group corresponding to each irreducible representation is given in Section 3.3. The notion of extended orbits can be used to further improve the computational efficiency.

We know that the block of each irreducible representation of the coordinate transformation matrix H for the symmetry of the Dihedral group has

$$\Sigma(H^{(1,1)}) = D_n, \qquad \Sigma(H^{(1,2)}) = C_n$$
$$\Sigma(H^{(1,3)}) = D_{n/2}, \qquad \Sigma(H^{(1,4)}) = D_{n/2}^2$$
$$\Sigma(H^{(2,j)+}) = D_{\gcd(j,n)}^k$$

$$\Sigma(H^{(2,j)-}) = \begin{cases} D_{\gcd(j,n)}^{k+n'}, & \text{when } n' = \text{even} \\ C_{\gcd(j,n)}, & \text{when } n' = \text{odd} \end{cases}$$

$$1 \le k \le n', \; j = 1,\cdots,m_d, \quad n' = n/[2\gcd(j,n)] \tag{5.24}$$

symmetry [30,31], where $\Sigma(\cdot)$ means the symmetry group of the column vector of the matrix in parentheses.

5.2.2 Matrix in Control System

Consider the equation of motion

$$\frac{d^k u}{dt^k} = \sum_{j=0}^{k-1} B_j \frac{d^j u}{dt^j} + \Gamma_k^{-1} f \tag{5.25}$$

Here, it is given by a certain k-order differential equation (see Eq.(3.2)).

$$B_j = -\Gamma_k^{-1}\Gamma_j \tag{5.26}$$

Defining a new variable

$$\widehat{u} = \left\{ u^{\mathrm{T}}, \left(\frac{du}{dt} \right)^{\mathrm{T}}, \cdots, \left(\frac{d^{k-1}u}{dt^{k-1}} \right)^{\mathrm{T}} \right\}^{\mathrm{T}}, \tag{5.27}$$

the differential equation of the control system denoted

$$\frac{d\widehat{u}}{dt} = A\widehat{u} + P \tag{5.28}$$

can be obtained from Eq.(5.25). Here, we define it as

$$A = \begin{pmatrix} O & I & O & \cdots & O \\ O & O & I & \cdots & O \\ \vdots & \vdots & \ddots & \ddots & \vdots \\ O & O & O & \ddots & I \\ B_0 & B_1 & B_2 & \cdots & B_{k-1} \end{pmatrix}, \tag{5.29}$$

$$P = \left\{ \mathbf{0}^{\mathrm{T}}, \mathbf{0}^{\mathrm{T}}, \cdots, (\Gamma_k^{-1} f)^{\mathrm{T}} \right\}^{\mathrm{T}} \tag{5.30}$$

Thus, the analysis of the control system requires an eigenvalue analysis of the kN-dimensional matrix A. It is clear from Eq.(5.29) that as matrix A is asymmetric and has a large bandwidth.

　　Next, we define

$$\widehat{w}^{\mu} = \left\{ (w^{\mu})^{\mathrm{T}}, \left(\frac{dw^{\mu}}{dt} \right)^{\mathrm{T}}, \cdots, \left(\frac{d^{k-1}w^{\mu}}{dt^{k-1}} \right)^{\mathrm{T}} \right\}^{\mathrm{T}}, \quad {}^{\forall}\mu \in R(G) \tag{5.31}$$

as the variable corresponding to the irreducible representation for this variable $\widehat{w}^{\mu}$, we can directly derive the expression corresponding to the expression (5.28) from the expression (3.13) to obtain an independent expression for each irreducible space of

$$\frac{d\widehat{w}^{\mu}}{dt} = \widetilde{A}^{\mu}\widehat{w}^{\mu} + \widetilde{P}^{\mu}, \quad {}^{\forall}\mu \in R(G) \tag{5.32}$$

where

$$\widetilde{A}^{\mu} = \begin{pmatrix} O & I & O & \cdots & O \\ O & O & I & \cdots & O \\ \vdots & \vdots & \ddots & \ddots & \vdots \\ O & O & O & \ddots & I \\ \widetilde{B}_0^{\mu} & \widetilde{B}_1^{\mu} & \widetilde{B}_2^{\mu} & \cdots & \widetilde{B}_{k-1}^{\mu} \end{pmatrix}, \tag{5.33}$$

$$\widetilde{P}^{\mu} = \left\{ \mathbf{0}^{\mathrm{T}}, \mathbf{0}^{\mathrm{T}}, \cdots, [(\widetilde{\Gamma}_k^{\mu})^{-1}(H^{\mu})^{\mathrm{T}} f]^{\mathrm{T}} \right\}^{\mathrm{T}} \tag{5.34}$$

and the matrix $\tilde{B}_j^\mu$ is (see Eq.(5.26))

$$
\begin{aligned}
\tilde{B}_j^\mu &= -(\tilde{\Gamma}_k^\mu)^{-1}\tilde{\Gamma}_j^\mu \\
&= -\left((H^\mu)^{\mathrm{T}}\Gamma_k H^\mu\right)^{-1}\left((H^\mu)^{\mathrm{T}}\Gamma_j H^\mu\right) \\
&= -(H^\mu)^{-1}\Gamma_k^{-1}[(H^\mu)^{\mathrm{T}}]^{-1}(H^\mu)^{\mathrm{T}}\Gamma_j H^\mu \\
&= -(H^\mu)^{\mathrm{T}}\Gamma_k^{-1}\Gamma_j H^\mu \\
&= (H^\mu)^{\mathrm{T}}B_j H^\mu
\end{aligned}
\tag{5.35}
$$

The solution of the differential equation of the control system (5.25) is obtained by superposing the solution $\widehat{w}^\mu$ of Eq.(5.32) over Eq.(5.20). Also, B_j can be block diagonalized with

$$
\tilde{B}_j = H^{\mathrm{T}}B_j H = \mathrm{diag}[\cdots, \tilde{B}_j^\mu, \cdots]
\tag{5.36}
$$

using $N \times N$ coordinate transformation matrix H directly. Thus, using this method, eigenvalue analysis of an asymmetric matrix A can be replaced by eigenvalue analysis of multiple small-sized matrices $\tilde{A}^\mu$, thereby improving the efficiency of the numerical analysis.

5.3 Symmetrical Transformation of Translational Displacement and Rotational Displacement

The translational displacement vector v^i of the node $i(i = 1, 2, \cdots)$ is shown in **Fig. 5.1**. Let us consider a generalized displacement vector

$$
u = \left\{ \begin{array}{c} v \\ \theta \end{array} \right\} = \left\{ \begin{array}{c} [\cdots, (v^i)^{\mathrm{T}}, \cdots]^{\mathrm{T}} \\ [\cdots, (\theta^i)^{\mathrm{T}}, \cdots]^{\mathrm{T}} \end{array} \right\}
\tag{5.37}
$$

consisting of a rotational displacement vector θ^i . Here,

$$
v^i = \{v_X^i, v_Y^i, v_Z^i\}^{\mathrm{T}}, \quad \theta^i = \{\theta_X^i, \theta_Y^i, \theta_Z^i\}^{\mathrm{T}}
\tag{5.38}
$$

In this figure, $\rightarrow$ represents the translational displacement in that direction and $\twoheadrightarrow$ represents the rotational displacement around that axis. Since the D_n-invariant structural system has rotational symmetry around the Z-axis, the displacements in the X, Y, and Z directions have different properties. Correspondingly, let v^i and θ^i be decomposed into

$$
v^i = \left\{ \begin{array}{c} v_{XY}^i \\ v_Z^i \end{array} \right\}, \quad \theta^i = \left\{ \begin{array}{c} \theta_{XY}^i \\ \theta_Z^i \end{array} \right\}
\tag{5.39}
$$

respectively. Where the subscript XY (or Z) of each vector represents the component corresponding to the X, Y (or Z) direction.

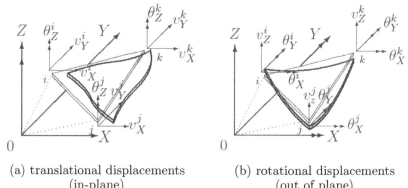

(a) translational displacements   (b) rotational displacements
          (in-plane)   (out of plane)

FIGURE 5.1
Displacement definition for a triangular plate

Translational displacement v^i and rotational displacement θ^i are independent of each other, we have the representation matrix $\widehat{T}(g)$ of the displacements with 6 degrees of freedom at one node;

$$\widehat{T}(g) = \begin{pmatrix} \widehat{T}_v(g) & O \\ O & \widehat{T}_\theta(g) \end{pmatrix}, \quad g \in G \tag{5.40}$$

which can be expressed in the form of a direct sum of the representation matrix $\widehat{T}_v(g)$ for translational displacements and $\widehat{T}_\theta(g)$ for rotational displacements.[1]

It is clear from **Fig. 5.2** that the rotation operation r has the same action on translational and rotational displacements, so the relation

$$\widehat{T}_\theta(r) = \widehat{T}_v(r) \tag{5.41}$$

is established for both representation matrices.

The translational displacement vector of a node is transformed for a mirror transformation s as shown in **Fig. 5.3**(a). On the other hand, the rotational displacement vector is in the opposite direction of the translational displacement direction, as shown in **Fig. 5.3**(b). This is because translational displacement is a polar vector, whereas rotational displacement is an axial vector. These results can be summarized as follows;

$$\widehat{T}_\theta(s) = -\widehat{T}_v(s) = \begin{pmatrix} 1 & 0 & 0 \\ 0 & -1 & 0 \\ 0 & 0 & -1 \end{pmatrix} \tag{5.42}$$

---

[1]The representation matrix for a single orbit is $\widehat{T}(g)$ and is the tensor product of the substitution representation and the direct sum over all those orbits is equal to $T(g)$. See Ikeda-Murota [27, 28] for details.

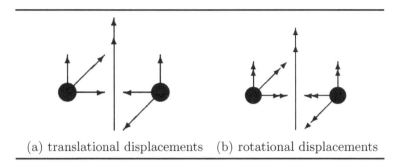

(a) translational displacements   (b) rotational displacements

FIGURE 5.2
Effect of rotational transformation r on displacement vector

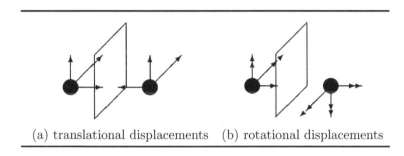

(a) translational displacements   (b) rotational displacements

FIGURE 5.3
Effect of mirror transformation s on displacement vector

The relation that the representation matrices $\widehat{T}_\theta(s)$ and $\widehat{T}_v(s)$ are opposite in sign and have the same components is called the difference in the action of the mirror transformation on translational and rotational displacements.

5.3.1 Interrelationship between Translational Displacement and Rotational Displacement Transformation Matrix

Coordinate transformation matrix $H = H_v$ for the representation matrix

$$T(g) = T_v(g) \tag{5.43}$$

of translational displacements is obtained by Murota-Ikeda [27, 28]. We note here that the representation matrix

$$T(g) = \begin{pmatrix} T_v(g) & O \\ O & T_\theta(g) \end{pmatrix}, \quad g \in G \tag{5.44}$$

TABLE 5.1
Relationship between the irreducible representations μ and $\bar{\mu}$

1st ordered irreducible representation		2nd ordered irreducible representation	
μ	$\bar{\mu}$	μ	$\bar{\mu}$
$(1,1)$	$(1,2)$	$(2,1)\pm$	$(2,1)\mp$
$(1,2)$	$(1,1)$	$(2,2)\pm$	$(2,2)\mp$
		$\vdots$	$\vdots$
$(1,3)$	$(1,4)$		
$(1,4)$	$(1,3)$	$(2,m_2)\pm$	$(2,m_2)\mp$

for the generalized displacement $\boldsymbol{u}$ and the coordinate transformation matrix

$$H^{\mu} = \begin{pmatrix} H^{\mu}_v & O \\ O & H^{\mu}_{\theta} \end{pmatrix}, \quad \mu \in R(G) \tag{5.45}$$

for each irreducible representation are in the form of a direct sum.

Writing Eq.(2.61) for $\boldsymbol{v}$ and $\boldsymbol{\theta}$, we get

$$H^{\mathrm{T}}_{\Phi} T_{\Phi}(g) H_{\Phi} = \bigoplus_{\mu \in R(G)} \overset{a^{\mu}_{\Phi}}{\bigoplus_{i=1}} T^{\mu}_i(g), \quad g \in G, \quad \Phi = v \text{ or } \theta$$

where a^{μ}_v is the representation matrix $T_v(g)$ and a^{μ}_{θ} denotes the degree of overlap of the irreducible representation μ in the representation matrix $T_{\theta}(g)$, respectively. Here, we clarify the relationship between A^{μ}_v and A^{μ}_{θ}, and between H^{μ}_v and H^{μ}_{θ}, and show that based on the concrete forms of a^{μ}_v and H^{μ}_v obtained by Murota-Ikeda [27, 28], we shall obtain a^{μ}_{θ} and H^{μ}_{θ}.

The expression matrix relations (5.41) and (5.42) for the displacement $\boldsymbol{u}^i$ of one node are held for all nodal displacements $\boldsymbol{u}$ and can be summarized as

$$T_{\theta}(g) = \sigma(g) T_v(g), \quad g \in G \tag{5.46}$$

$$\sigma(g) = \begin{cases} 1, & g = r \\ -1, & g = s \end{cases} \tag{5.47}$$

On the other hand, for an irreducible representation μ, the irreducible representation $\bar{\mu}$ is defining **Table 5.1**, so we obtain the relation

$$\sigma(g) T^{\mu}(g) = T^{\bar{\mu}}(g), \quad g \in G \tag{5.48}$$

From Eqs. (5.46) and (5.48) the following relationship between multiplicity and the oordinate transformation matrix can be obtained. The translational displacement $\boldsymbol{v}$ and the rotational displacement $\boldsymbol{\theta}$ have the same multiplicity and

$$a^{\mu}_{\theta} = a^{\mu}_v \tag{5.49}$$

holds.

The relation shown in **Table 5.1** is established for the irreducible representations μ and $\bar{\mu}$ of the translational displacement $\boldsymbol{v}$ and rotational displacement $\boldsymbol{\theta}$. Adding considerations such as equations (5.24) and (5.47) to this relationship,

$$H_\theta^{(1,1)} = H_v^{(1,2)}, \qquad H_\theta^{(1,2)} = H_v^{(1,1)} \tag{5.50}$$

$$H_\theta^{(1,3)} = H_v^{(1,4)}, \qquad H_\theta^{(1,4)} = H_v^{(1,3)} \tag{5.51}$$

$$H_\theta^{(2,j)+} = H_v^{(2,j)-}, \qquad H_\theta^{(2,j)-} = H_v^{(2,j)+} \tag{5.52}$$

$$j = 1, \cdots, m_2$$

is obtained. The Eq.(5.50) allows us to calculate the rotational displacement H_θ^μ based on the translational displacement H_v^μ. Furthermore, based on H_v^μ and H_θ^μ obtained in this way, the coordinate transformation matrix H can be calculated from the expressions (5.16) and (5.45).

5.4 Orbit

5.4.1 Orbital Concept and Matrix Coordinate Transforms

The set of nodes of a D_n-invariant discretized structural system is also a D_n-invariant and can be decomposed into the smallest unit of (D_n-invariant) orbits. This trajectory is the same as the orbit of by the elements r^k and sr^k ($k = 0, 1, \cdots, n-1$) of D_n, the set of points $r^k(\boldsymbol{x})$ and $sr^k(\boldsymbol{x})$ transformed from a node $\boldsymbol{x}$, which can be defined as

$$\{r^k(\boldsymbol{x}),\ sr^k(\boldsymbol{x}) \in \mathbf{R}^2 \mid k = 0, 1, \cdots, n-1\} \tag{5.53}$$

Because of the coordinate transformations induced by D_n, the orbit as a whole has the property of remaining invariant even when individual nodes are shifted. D_n-invariant discrete system nodal orbits can be classified into four types,

$$\text{type of orbits} \begin{cases} \text{Center type} & (0) \\ n - \text{gon type} & (1\text{V}) \\ n - \text{gon type} & (1\text{M}) \\ 2n - \text{gon type} & (2) \end{cases}$$

as in **Fig. 3.1** [27, 31]. Determining the column vector of the matrix H of coordinate transformations for each orbit allows the matrix H to be taken sparsely, which is computationally more efficient. It also has the advantage that the computation of H can be handled systematically.

The H subblock matrix [2] of a structural system consisting of N_O orbits

---

[2]On the right-hand side of Eq.(5.50), Murota and Ikeda [27,28] et al. proposed substituting the formula for calculating H_v orbit-by-orbit, and the coordinate transformation matrix of rotational displacements H_θ is obtained.

is very sparse, of the form

$$H^\mu = \text{diag}[H_1^\mu, \cdots, H_{N_O}^\mu]$$

$$= \begin{pmatrix} H_1^\mu & O & O & O & O \\ O & \cdot & O & O & O \\ O & O & \cdot & O & O \\ O & O & O & \cdot & O \\ O & O & O & O & H_{N_O}^\mu \end{pmatrix}, \quad \mu \in R(G) \quad (5.54)$$

which has only one component for each orbital. Note that it is clear from Eq.(5.45) that the representation matrices of the translational displacement v and the rotational displacement θ are independent, and the representation matrices of the X, Y, and Z direction components of both matrices are also independent. Corresponding to this independence of the representation matrices, each block H_q^μ ($q = 1, \cdots, N_O$) of a coordinate transformation matrix. Furthermore, it has a sparse detailed block structure called

$$H_q^\mu = \text{diag}[H_{q,vXY}^\mu, H_{q,vZ}^\mu, H_{q,\theta XY}^\mu, H_{q,\theta Z}^\mu]$$

$$= \begin{pmatrix} H_{q,vXY}^\mu & O & O & O \\ O & H_{q,vZ}^\mu & O & O \\ O & O & H_{q,\theta XY}^\mu & O \\ O & O & O & H_{q,\theta Z}^\mu \end{pmatrix}, \quad \mu \in R(G) \quad (5.55)$$

The block matrix $\widetilde{K}^\mu$ of the stiffness matrix can be expressed by the formula (5.19). In computing the H matrix sparsity, such as in equations (5.54) and (5.55), it is highly advantageous to use the sparsity of the H matrix, such as in equations (5.54) and (5.55). For an efficient method of computing the block matrix $\widetilde{K}^\mu$ of the stiffness matrix, see the authors' reference [31].

5.4.2   Orbital Concept for Elements

D_n invariant system configurations and the concept of orbits can also be applied to elements. The trajectories for the elements are decomposed into four types; (1) Center type, (2)n-gon type I, (3) n-gon type II, and (4) $2n$-gon type (octagon) as shown in **Fig. 5.4**. The subgroup $G_{\text{sub}} \equiv \Sigma(H^\mu)$ of D_n belongs to one of these four orbits.

Two elements, e and e^*, belong to the same orbit between elements under the action of the subgroup G_{sub} and the nodes p_k and p_k^* (respectively, p_l and p_l^*) of these two elements belonging to the same orbit between nodes under the action of the same group G, respectively. In this case, there exists an irreducible representation matrix $T(g)$ for the source g of the subgroup G, and the relation

$$T(g)H_{q_\eta p_\eta}^\mu = H_{q_\eta^* p_\eta^*}^\mu, \quad \eta = k \text{ or } l, \quad (5.56)$$

$$\Gamma_{p_k^* p_l^*}^{e^*} = T(g)^{\text{T}} \Gamma_{p_k p_l}^e T(g), \quad g \in G \quad (5.57)$$

TABLE 5.2

For example, the number of elements belonging to the orbit extended between elements for $n = 4$

Type of Orbit	μ			
	$(1,1)_{D_4}, (1,2)_{D_4}, (1,3)_{D_4}, (1,4)_{D_4}$	$(2,1)^+_{D_4}$	$(2,1)^-_{D_4}$
Center	1	1	1
Square type I	4	4	2
Square type II	4	2	4
Octagonal	8	4	4

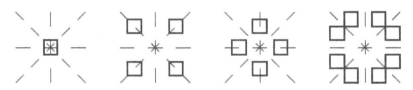

Center type   Square type I   Square type II Octagonal type

FIGURE 5.4

Orbit for elements

holds, where $\Gamma^{e^*}_{p^*_k p^*_l}$ is the p^*_k and p^*_l associated with the two points. It is a submatrix to the element stiffness matrix of element e^*.

For the treatment of equations (5.56) and (5.57), corresponding to each orbit and further decomposed in the irreducible representation μ, we have $\Gamma^e_{p_k p_l}$, which can be expressed by

$$
\begin{aligned}
\tilde{\Gamma}^{e\mu}_{q_k q_l} &\equiv \left(H^\mu_{q_k p_k}\right)^{\mathrm{T}} \Gamma^e_{p_k p_l} H^\mu_{q_l p_l} \\
&= \left(H^\mu_{q_k p_k}\right)^{\mathrm{T}} T(g)^{\mathrm{T}} T(g) \Gamma^e_{p_k p_l} T(g)^{\mathrm{T}} T(g) H^\mu_{q_l p_l} \\
&= \left[T(g) H^\mu_{q_k p_k}\right]^{\mathrm{T}} \left[T(g)^{\mathrm{T}} \Gamma^e_{p_k p_l} T(g)\right]^{\mathrm{T}} \left[T(g) H^\mu_{q_l p_l}\right] \\
&= \left(H^\mu_{q^*_k p^*_k}\right)^{\mathrm{T}} \Gamma^{e^*}_{p^*_k p^*_l} H^\mu_{q^*_l p^*_l} \\
&\equiv \tilde{\Gamma}^{e\mu}_{q^*_k q^*_l}
\end{aligned}
\tag{5.58}
$$

Equation (5.58) shows that elements belonging to the same orbit under the action of G have the same element stiffness as the group G.

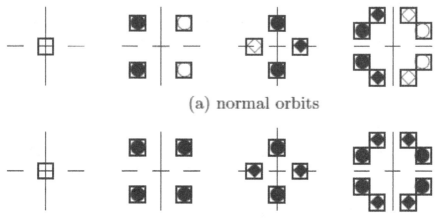

(a) normal orbits

(b) augmented orbits

FIGURE 5.5
Mirror symmetry for each orbit

Further extending the concept of trajectory;

$$T(g)H^\mu_{q_\eta p_\eta} = -H^\mu_{q^*_\eta p^*_\eta}, \quad \eta = k \text{ or } l, \quad g \in G. \tag{5.59}$$

This is similar to the expression (5.58) The extended invariant orbit formula (5.56) or (5.59) The equation is for nodes satisfying either of the following, that is, the conversion is equivalent to

$$
\begin{aligned}
\widetilde{\Gamma}^{e\mu}_{q_k q_l} &= \left[T(g)H^\mu_{q_k p_k} \right]^{\mathrm{T}} \Gamma^{e^*}_{p^*_k p^*_l} \left[T(g)H^\mu_{q_l p_l} \right] \\
&= \left[-T(g)H^\mu_{q_k p_k} \right]^{\mathrm{T}} \Gamma^{e^*}_{p^*_k p^*_l} \left[-T(g)H^\mu_{q_l p_l} \right] \\
&= \left(-H^\mu_{q^*_k p^*_k} \right)^{\mathrm{T}} \Gamma^{e^*}_{p^*_k p^*_l} \left(-H^\mu_{q^*_l p^*_l} \right) \\
&= \widetilde{\Gamma}^{e\mu}_{q^*_k q^*_l} \tag{5.60}
\end{aligned}
$$

from Eq.(5.58). Since elements with the same orbit extended have the same properties, the element stiffness between these elements must be calculated by the number of elements between orbits extended to the properties of the entire orbit. However, since the deformation pattern of each orbit is known, analysis of only some elements can be equivalent to analysis of the entire orbit, thus improving computational efficiency.

5.4.3   Local Coordinate Transformation of Various Matrices

The column vector H^μ is related to the Cartesian displacement vector $\boldsymbol{u}$ and the displacement vector $\boldsymbol{w}^\mu$ after decomposition of the irreducible representation μ. The components of $\boldsymbol{w}^\mu$ are the superposition of the deformation modes of each orbit, which can be decomposed into each component $\boldsymbol{w}_i^\mu$ of the orbit.

$$\boldsymbol{w}^\mu = \{(\boldsymbol{w}_1^\mu)^{\mathrm{T}}, \cdots, (\boldsymbol{w}_{N_\mathrm{O}}^\mu)^{\mathrm{T}}\}^{\mathrm{T}} \tag{5.61}$$

where. $N_\mathrm{O} = N_\mathrm{O}(\mu)$ is the number of orbits and $\boldsymbol{w}_i^\mu$ is the ith orbit ($i = 1, \cdots, N_\mathrm{O}$). Also, $\boldsymbol{u}$ is decomposed into nodal components;

$$\boldsymbol{u} = \{(\boldsymbol{u}_1)^{\mathrm{T}}, \cdots, (\boldsymbol{u}_{N_\mathrm{P}})^{\mathrm{T}}\}^{\mathrm{T}} \tag{5.62}$$

where N_P denote the number of nodes and $\boldsymbol{u}_i (i = 1, \cdots, N_\mathrm{P})$ denotes the i-th nodal displacement vector.

In the variables decomposed into equations (5.61) and (5.62), we have column vector H^μ and diagonal block $\widetilde{\Gamma}(= \widetilde{\Gamma}_j)$. The matrix $\Gamma(= \Gamma_j)$ is then partitioned into matrix blocks as follows;

$$
\begin{aligned}
H^\mu &= (H_{ij}^\mu \mid i = 1, \cdots, N_\mathrm{P}; j = 1, \cdots, N_\mathrm{O}) \\
&= \begin{pmatrix} H_{1,1}^\mu & \cdots & H_{1,N_\mathrm{O}}^\mu \\ \cdot & & \cdot \\ \cdot & & \cdot \\ H_{N_\mathrm{P},1}^\mu & \cdots & H_{N_\mathrm{P},N_\mathrm{O}}^\mu \end{pmatrix},
\end{aligned} \tag{5.63}
$$

$$\Gamma^\mu = (\Gamma_{ij}^\mu \mid i, j = 1, \cdots, N_\mathrm{O}), \tag{5.64}$$

$$\Gamma = (\Gamma_{ij} \mid i, j = 1, \cdots, N_\mathrm{P}). \tag{5.65}$$

Denote the orbit number of each node by

$$(q_i \equiv q_i(p_i) \mid i = 1, \cdots, M)$$

and The set of nodes with node numbers M ($p_i \mid i = 1, \cdots, M$), and consider a board consisting of N_e elements of the eth element ($e = 1, \cdots, N_e$). The e-th matrix Γ^e ($e = 1, \cdots, N_e$) can be split into

$$\Gamma^e = (\Gamma_{p_i}^e \boldsymbol{w}^{(2,j)-} \mid i, j = 1, \cdots, M). \tag{5.66}$$

and matrix blocks.

In this case, the operation of the expression (5.17) on the entire structure must be recombined with the operation on

$$\Gamma_{ij}^\mu = \sum_{e=1}^{N_e} \sum_{k=1}^{M} \sum_{l=1}^{M} \Gamma_{q_k q_l}^{e\mu} \delta_{iq_k} \delta_{jq_l}, \tag{5.67}$$

and the elements. Where δ_{iq_k} and δ_{jq_l} denote Kronecker's delta and

$$\Gamma^{e\mu}_{q_k q_l} = \left(H^{\mu}_{q_k p_k}\right)^{\mathrm{T}} \Gamma^{e}_{p_k p_l} H^{\mu}_{q_l p_l}. \tag{5.68}$$

respectively. If Γ^{μ} is a symmetric matrix, then it is sufficient to compute Γ^{μ}_{ij} with $i \geq j$. Also, $\widetilde{\Gamma}^{(2,1)+}$ and $\widetilde{\Gamma}^{(2,1)-}$ are the same, so we can use the condition that The computational cost of Eq.(5.67) can be further reduced. Since the matrix size of the right-hand side of Eq.(5.68) is significantly smaller than that of Eq.(5.17), the cost for the matrix operation is significantly reduced.

6

Numerical Analysis of Dihedral Group Invariant Systems

In this chapter, static and dynamic problems for D_n-invariant structures with geometrical symmetry are addressed. The numerical calculations are discussed, together with the technical calculation methods involved in the compatibility of this theory with the finite element method. Accordingly, a comparative study of conventional structural calculation time and required array capacity for symmetrical structures was conducted. As a result, significant improvement in computing efficiency was achieved. The results of the analysis of the efficiency of the calculation were also obtained.

6.1 Element Stiffness for Static Analysis

6.1.1 Assembly of Coordinate Transformation Matrix

This chapter describes a method for assembling the coordinate transformation matrix H for D_n-invariant plates using the concept of orbits (see reference [19]). The D_n-invariance also holds at the nodes for the whole D_n-invariant plate. All nodes are partitioned as D_n-invariant subsets. This is called the **orbit** in mathematics. The sources r^k and sr^k ($k = 0, \cdots, n-1$) of D_n are transformed by nodes $\boldsymbol{x}$ to $r^k\boldsymbol{x}$, $(sr^k)\boldsymbol{x}$, respectively. This method defines the orbits of the nodes.

$$\{r^k\boldsymbol{x} \text{ and } (sr^k)\boldsymbol{x} \mid k = 0, \cdots, n-1\} \tag{6.1}$$

After the transformation, it is transformed to another coordinate, but the invariance of the plate itself is preserved due to the original geometric transformation of D_n. As seen in **Fig. 6.1**, there are four types of orbits The four types are (1) Center type, (2) n-gon type I, (3) n-gon type II, and (4) $2n$-gon type. In particular, for square plates ($n = 4$) the n-gon type is called the Square type and the $2n$-gon type is called the Octagonal type. For example, when the square plate in **Fig. 6.2** consists of one Center type, two Square type I, two Square type II, and one Octagonal type. The finite element method can be represented as an assembly structure of discretized plate elements as well as discretized plate elements as an assembly of trajectories.

DOI: 10.1201/9781032670386-6

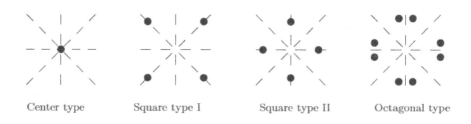

| Center type | Square type I | Square type II | Octagonal type |

FIGURE 6.1
Orbit types for rectangular nodes and elements

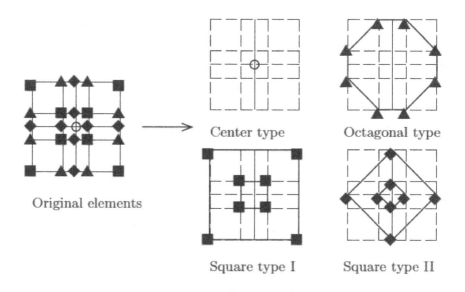

Center type

Octagonal type

Original elements

Square type I

Square type II

FIGURE 6.2
Orbital decomposition of square plate

The column vectors of the coordinate transformation matrix H correspond to the irreducible representation of each orbit. Such a definition not only creates a simpler formula, but also results in a sparser matrix H such that the components consisting of the orbits are non-zero. The orbital decomposition described in this book is more analytically efficient than that of Zloković (1989). This is because it decomposes more orbitals and H sparsity at the plate. For example, each contracted representation μ for $n = 4$ is related to a deformation mode such as the orbit of **Fig. 6.3**, where the solid line represents the deformed state and the dashed line represents the initial state. Some of

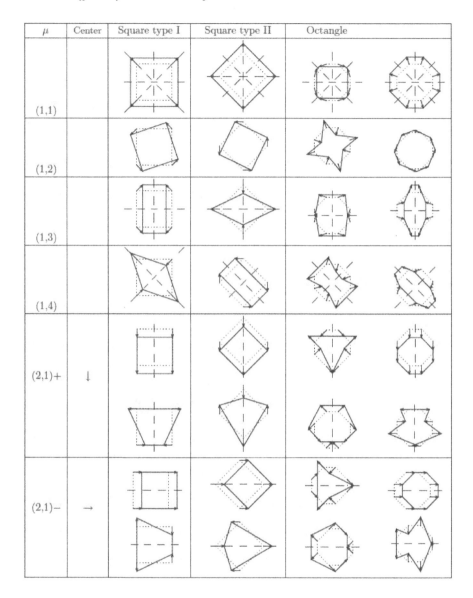

FIGURE 6.3
Transformation mode for each contracted representation

these deformation modes have rotational and mirror symmetry. For example,
the mode for $\mu = (1,2)$ has no mirror symmetry, however, they are related
to the four mirror symmetries of sr^k $(k = 0,1,2,3)$ that represent symmetry

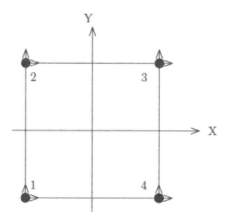

FIGURE 6.4
Coordinate system of 4 nodes

operations but are independent of their magnitude. Moreover, the mode corresponding to $\mu = (2,1)-$ contains symmetry operations corresponding to s and r^2 and is transformed exactly as the invariance corresponding to sr^2.

6.1.2 Diagonalization of Square Plate Plane Elements

As a simple example, let us assume an isotropic homogeneous material of uniform thickness t and consider a D_4 invariant square plate with constant Young's modulus E and Poisson's ratio ν. This plate is analyzed by a 4-node first-order element with 6 degrees of freedom as seen in **Fig. 6.4**, by the formula (3.23) for the element stiffness matrix K^e of the D_4-invariant plate. The transformation matrices $T(r^k)$ and $T(sr^k)$ $(k = 0, 1, 2, 3)$ for all elements of D_4 are obtained. For example, $T(r)$ and $T(s)$ become

$$
T(r) = \left(
\begin{array}{ccc|ccc}
 & & & & & -1 \\
 & & & & 1 & \\
 & -1 & & & & \\
1 & & & & & \\
 & & & -1 & & \\
 & & 1 & & & \\
 & & & & -1 & \\
 & & & 1 & & \\
\end{array}
\right),
$$

$$
T(s) = \left(
\begin{array}{cccc|cccc}
 & & & 1 & & & & \\
 & & -1 & & & & & \\
 & 1 & & & & & & \\
 -1 & & & & & & & \\
\hline
 & & & & 1 & & & \\
 & & & & & -1 & & \\
 & & & & & & 1 & \\
 & & & & & & & -1
\end{array}
\right).
$$

This condition (3.23) consists of K_σ in plane stress problem such that

$$
K_\sigma = \left(
\begin{array}{cccccccc}
B_1 & & & & & & & \\
B_4 & B_1 & & & \text{Symmetry} & & & \\
B_7 & -B_3 & B_1 & & & & & \\
B_3 & B_5 & -B_4 & B_1 & & & & \\
B_2 & -B_6 & B_5 & -B_3 & B_1 & & & \\
-B_6 & B_2 & B_3 & B_7 & B_4 & B_1 & & \\
B_5 & B_3 & B_2 & B_6 & B_7 & -B_3 & B_1 & \\
-B_3 & B_7 & B_6 & B_2 & B_3 & B_5 & -B_4 & B_1
\end{array}
\right)
\tag{6.2}
$$

where $(B_i \mid i = 1, \cdots, 7)$ is constant.

$$
B_i = \frac{Et}{12(1 - \nu^2)} b_i, \quad i = 1, \cdots, 7,
$$

$$
b_1 = 2(3 - \nu), \quad b_2 = -(3 - \nu), \quad b_3 = \frac{3}{2}(1 - 3\nu),
$$

$$
b_4 = b_6 = \frac{3}{2}(1 + \nu), \quad b_5 = -(3 + \nu), \quad b_7 = 2\nu.
$$

For example, the element stiffness matrix in plane stress problem for a Serendipity square element with four nodes forms Eq.(6.2). However, this element stiffness matrix is related to $T(g)$ and preserves the invariance of $T(g)$. As seen in **Fig. 6.3**, the expression (5.16) for the element transformation matrix H^μ consisting of Square type I orbits ($\mu = (1,1)_{D_4}, (1,2)_{D_4}, (1,3)_{D_4}, (1,4)_{D_4}, (2,1)^+_{D_4}, (2,1)^-_{D_4}$) gives the transformation pattern of this orbit.

The coordinate transformation matrix for the first-order irreducible representation is

$$
H^{(1,1)_{D_4}} = \begin{pmatrix}
-\sqrt{2}/4 \\
-\sqrt{2}/4 \\
-\sqrt{2}/4 \\
\sqrt{2}/4 \\
\sqrt{2}/4 \\
\sqrt{2}/4 \\
\sqrt{2}/4 \\
-\sqrt{2}/4
\end{pmatrix}, \quad
H^{(1,2)_{D_4}} = \begin{pmatrix}
\sqrt{2}/4 \\
-\sqrt{2}/4 \\
-\sqrt{2}/4 \\
-\sqrt{2}/4 \\
-\sqrt{2}/4 \\
\sqrt{2}/4 \\
\sqrt{2}/4 \\
\sqrt{2}/4
\end{pmatrix},
$$

$$H^{(1,3)_{D_4}} = \begin{pmatrix} -\sqrt{2}/4 \\ \sqrt{2}/4 \\ -\sqrt{2}/4 \\ -\sqrt{2}/4 \\ \sqrt{2}/4 \\ -\sqrt{2}/4 \\ \sqrt{2}/4 \\ \sqrt{2}/4 \end{pmatrix}, \quad H^{(1,4)_{D_4}} = \begin{pmatrix} -\sqrt{2}/4 \\ -\sqrt{2}/4 \\ \sqrt{2}/4 \\ -\sqrt{2}/4 \\ \sqrt{2}/4 \\ \sqrt{2}/4 \\ -\sqrt{2}/4 \\ \sqrt{2}/4 \end{pmatrix},$$

and the transformation matrix corresponding to the second-order irreducible representation is

$$H^{(2,1)^+_{D_4}} = \begin{pmatrix} 1/2 & 0 \\ 0 & -1/2 \\ 1/2 & 0 \\ 0 & 1/2 \\ 1/2 & 0 \\ 0 & -1/2 \\ 1/2 & 0 \\ 0 & 1/2 \end{pmatrix}, \quad H^{(2,1)^-_{D_4}} = \begin{pmatrix} 0 & -1/2 \\ 1/2 & 0 \\ 0 & 1/2 \\ 1/2 & 0 \\ 0 & -1/2 \\ 1/2 & 0 \\ 0 & 1/2 \\ 1/2 & 0 \end{pmatrix}.$$

At this time K_σ is transformed into a block diagonal with a square block in Eq.(5.17).

$$\widetilde{K}_\sigma \equiv H^T K_\sigma H$$

$$= \mathrm{diag}[\widetilde{K}_\sigma^{(1,1)_{D_4}}, \widetilde{K}_\sigma^{(1,2)_{D_4}}, \widetilde{K}_\sigma^{(1,3)_{D_4}}, \widetilde{K}_\sigma^{(1,4)_{D_4}}, \widetilde{K}_\sigma^{(2,1)^+_{D_4}}, \widetilde{K}_\sigma^{(2,1)^-_{D_4}}]$$

$$= \frac{Et}{1-\nu^2} \begin{bmatrix} \boxed{1+\nu} & & & & & & & \\ & 0 & & & & O & & \\ & & \boxed{1-\nu} & & & & & \\ & & & \boxed{1-\nu} & & & & \\ & & & & 0 & 0 & & \\ & & & & 0 & \frac{3-\nu}{6} & & \\ & & O & & & & 0 & 0 \\ & & & & & & 0 & \frac{3-\nu}{6} \end{bmatrix} \tag{6.3}$$

$$\widetilde{K}_\sigma^{(1,1)_{D_4}} = \frac{4Et}{1-\nu^2}\left(\frac{1}{8} + \frac{2\nu}{8} + \frac{1}{8}\right) = \frac{Et}{1-\nu}$$

$$\widetilde{K}_\sigma^{(1,2)_{D_4}} = 0$$

$$\widetilde{K}_\sigma^{(1,3)_{D_4}} = \widetilde{K}_\sigma^{(1,4)_{D_4}} = \frac{Et}{1+\nu}$$

$$\widetilde{K}_\sigma^{(2,1)^+_{D_4}} = \widetilde{K}_\sigma^{(2,1)^-_{D_4}} = \frac{Et}{1-\nu^2} \iint \begin{bmatrix} 0 & 0 \\ 0 & \frac{3-\nu}{2}\left(\frac{1}{2}\frac{\partial}{\partial\eta}\xi\eta\right)^2 \end{bmatrix} \det|J|\,d\xi\,d\eta$$

$$= \frac{Et(3-\nu)}{6(1-\nu^2)}\begin{pmatrix} 0 & 0 \\ 0 & 1 \end{pmatrix}$$

The three diagonal components of $\widetilde{K}_\sigma^{(1,2)_{D_4}}$, $\widetilde{K}_\sigma^{(2,1)_{D_4}^+}$, $\widetilde{K}_\sigma^{(2,1)_{D_4}^-}$ are zero in the rigid body displacement mode.

Question: If you have a rectangular plane stress problem with side lengths of $(2a \times 2b)$, consider similarly the transformed stiffness matrix. Then, it is available to use same transformation matrix H in D_4 invariant square plate.

Answer:

$$\widetilde{K}_\sigma = \mathrm{diag}[\widetilde{K}_\sigma^{(1,1)_{D_2}}, \widetilde{K}_\sigma^{(1,3)_{D_2}}, \widetilde{K}_\sigma^{(2,1)_{D_2}^+}, \widetilde{K}_\sigma^{(2,1)_{D_2}^-}]$$

$$= \frac{Et}{1-\nu^2} \left[\begin{pmatrix} \frac{1}{2}\left(\frac{a}{b} + \frac{b}{a}\right) + \nu & \frac{1}{2}\left(\frac{b}{a} - \frac{a}{b}\right) \\ \frac{1}{2}\left(\frac{b}{a} - \frac{a}{b}\right) & \frac{1}{2}\left(\frac{a}{b} + \frac{b}{a}\right) - \nu \end{pmatrix} \right.$$

$$\oplus \frac{1-\nu}{4ab} \begin{pmatrix} (a-b)^2 & -(a-b)(a+b) \\ -(a-b)(a+b) & (a+b)^2 \end{pmatrix}$$

$$\left. \oplus \begin{pmatrix} 0 & 0 \\ 0 & \frac{2a^2 + b^2(1-\nu)}{6ab} \end{pmatrix} \oplus \begin{pmatrix} 0 & 0 \\ 0 & \frac{a^2(1-\nu) + 2b^2}{6ab} \end{pmatrix} \right]$$

The stiffness matrix form of this block diagonalization is shown in **Fig. 6.5**. This is related to the irreducible representations in D_4 group in the following;

Square is D_4 symmetry : Rectangle is D_2 symmetry

$$\left. \begin{matrix} (1,1)_{D_4} \\ (1,2)_{D_4} \end{matrix} \right\} \rightarrow (1,1)_{D_2}$$

$$\left. \begin{matrix} (1,3)_{D_4} \\ (1,4)_{D_4} \end{matrix} \right\} \rightarrow (1,3)_{D_2}$$

$$(2,1)_{D_4}^+ \rightarrow (2,1)_{D_2}^+$$

$$(2,1)_{D_4}^- \rightarrow (2,1)_{D_2}^-$$

Question: How is the block diagonalization of the stiffness matrix for a square element plane strain problem transformed?

Answer: It is finally obtained the following matrix form;

$$\widetilde{K}_e^\varepsilon = \frac{Et}{(1+\nu)(1-2\nu)} \begin{bmatrix} 1 & & & & & \\ & 0 & & & O & \\ & & 1-2\nu & & & \\ & & & 1-2\nu & & \\ & & & & 0 & 0 \\ & & & & 0 & \frac{1}{6}(3-4\nu) \\ & O & & & & 0 & 0 \\ & & & & & 0 & \frac{1}{6}(3-4\nu) \end{bmatrix} \tag{6.4}$$

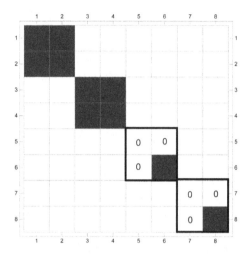

FIGURE 6.5
Block diagonalization matrix form for a rectangle plate (brown for non-zero components, white for zero components)

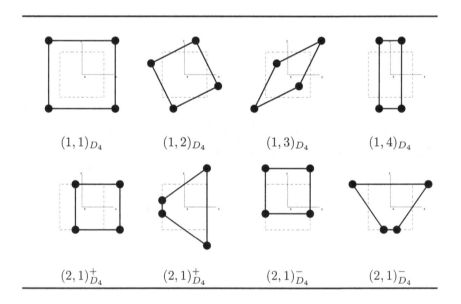

FIGURE 6.6
XY−plane deformations for each irreducible representation of D_4

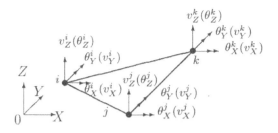

FIGURE 6.7
Equilateral triangle plate element

6.1.3 D_3 Block Diagonalization of the Stiffness Matrix of the Immutable Plate Element

Nodal displacements with 6 degrees of freedom at each node (3 degrees of translational displacement and 3 degrees of rotational displacement). The element stiffness matrix of a three-node element of an equilateral triangular plate is block-diagonalized by this method.

Although there is no unique way to take a coordinate system, here, the nodes of the element are taken to form a 1M type orbit, as shown in **Fig. 6.7**. The expression (5.16) of the coordinate transformation matrix for this element is expressed as

$$H = [H^{(1,1)_{D_3}}, H^{(1,2)_{D_3}}, H^{(2,1)_{D_3}^+}, H^{(2,1)_{D_3}^-}] \qquad (6.5)$$

Each column vector of the coordinate transformation matrix is shown in **Fig. 6.3** given by the transformation mode. The concrete form of the coordinate transformation matrix is shown here in **Fig. 6.9**. This H matrix is very sparse as is evident from **Fig. 6.3**. This sparsity is due to the double block structure shown in equation (5.16). In a numerical analysis program, there is no need to define the H-matrix itself; using only the information about the symmetric subgroup (trajectory) of each irreducible representation as shown in **Fig. 6.8** will reduce storage space and computational cost.

The element stiffness matrix K_e of this equilateral triangular plate is **Fig. 6.10**. Substituting the respective matrix components into the transformation formula

$$\begin{aligned} \widetilde{K}_e &= H^T K_e H \\ &= \mathrm{diag}[\widetilde{K}_e^{(1,1)}, \widetilde{K}_e^{(1,2)}, \widetilde{K}_e^{(2,1)+}, \widetilde{K}_e^{(2,1)-}] \end{aligned} \qquad (6.6)$$

results in a block diagonalization as shown in **Fig. 6.11**.

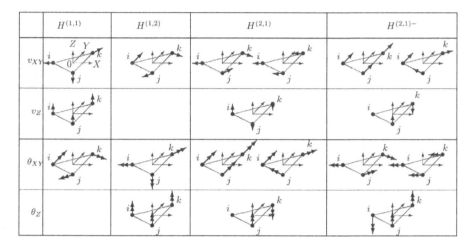

FIGURE 6.8
Deformation pattern represented by column vector H (corresponding to 1M type)

	$H^{(1,1)}$			$H^{(1,2)}$			$H^{(2,1)}$						$H^{(2,1)-}$					
	v_{XY}	v_Z	θ_{XY}	v_{XY}	θ_{XY}	θ_Z	v_{XY}	v_{XY}	v_Z	θ_{XY}	θ_{XY}	θ_Z	v_{XY}	v_{XY}	v_Z	θ_{XY}	θ_{XY}	θ_Z
v_X^i	$\frac{-1}{\sqrt{3}}$	0	0	0	0	0	$\frac{-1}{\sqrt{3}}$	$\frac{-1}{\sqrt{3}}$	0	0	0	0	0	0	0	0	0	0
v_Y^i	0	0	0	$\frac{1}{\sqrt{3}}$	0	0	0	0	0	0	0	0	$\frac{1}{\sqrt{3}}$	$\frac{-1}{\sqrt{3}}$	0	0	0	0
θ_Z^i	0	0	0	0	0	$\frac{1}{\sqrt{3}}$	0	0	0	0	0	0	0	0	0	0	0	$\frac{2}{\sqrt{6}}$
v_Z^i	0	$\frac{1}{\sqrt{3}}$	0	0	0	0	0	0	$\frac{2}{\sqrt{6}}$	0	0	0	0	0	0	0	0	0
θ_X^i	0	0	0	0	$\frac{-1}{\sqrt{3}}$	0	0	0	0	0	0	0	0	0	0	$\frac{1}{\sqrt{3}}$	$\frac{-1}{\sqrt{3}}$	0
θ_Y^i	0	0	$\frac{1}{\sqrt{3}}$	0	0	0	0	0	0	$\frac{1}{\sqrt{3}}$	$\frac{-1}{\sqrt{3}}$	0	0	0	0	0	0	0
v_X^j	$\frac{1}{2\sqrt{3}}$	0	0	$\frac{-1}{2}$	0	0	$\frac{1}{\sqrt{3}}$	$\frac{1}{2\sqrt{3}}$	0	0	0	0	0	$\frac{-1}{2}$	0	0	0	0
v_Y^j	$\frac{-1}{2}$	0	0	$\frac{-1}{2\sqrt{3}}$	0	0	0	$\frac{1}{2}$	0	0	0	0	$\frac{1}{\sqrt{3}}$	$\frac{1}{2\sqrt{3}}$	0	0	0	0
θ_Z^j	0	0	0	0	0	$\frac{1}{\sqrt{3}}$	0	0	0	0	0	$\frac{-1}{\sqrt{2}}$	0	0	0	0	0	$\frac{-1}{\sqrt{6}}$
v_Z^j	0	$\frac{1}{\sqrt{3}}$	0	0	0	0	0	0	$\frac{-1}{\sqrt{6}}$	0	0	0	0	0	$\frac{-1}{\sqrt{2}}$	0	0	0
θ_X^j	0	0	$\frac{-1}{2}$	0	$\frac{1}{2\sqrt{3}}$	0	0	0	0	$\frac{-1}{2}$	0	0	0	0	0	$\frac{1}{\sqrt{3}}$	$\frac{1}{2\sqrt{3}}$	0
θ_Y^j	0	0	$\frac{-1}{2\sqrt{3}}$	0	$\frac{-1}{2}$	0	0	0	0	$\frac{1}{\sqrt{3}}$	$\frac{1}{2\sqrt{3}}$	0	0	0	0	$\frac{1}{2}$	0	0
v_X^k	$\frac{1}{2\sqrt{3}}$	0	0	$\frac{1}{2}$	0	0	$\frac{1}{\sqrt{3}}$	$\frac{1}{2\sqrt{3}}$	0	0	0	0	0	$\frac{1}{2}$	0	0	0	0
v_Y^k	$\frac{1}{2}$	0	0	$\frac{-1}{2\sqrt{3}}$	0	0	0	$\frac{-1}{2}$	0	0	0	0	$\frac{1}{\sqrt{3}}$	$\frac{1}{2\sqrt{3}}$	0	0	0	0
θ_Z^k	0	0	0	0	0	$\frac{1}{\sqrt{3}}$	0	0	0	0	0	$\frac{1}{\sqrt{2}}$	0	0	0	0	0	$\frac{-1}{\sqrt{6}}$
v_Z^k	0	$\frac{1}{\sqrt{3}}$	0	0	0	0	0	0	$\frac{-1}{\sqrt{6}}$	0	0	0	0	0	$\frac{1}{\sqrt{2}}$	0	0	0
θ_X^k	0	0	$\frac{1}{2}$	0	$\frac{1}{2\sqrt{3}}$	0	0	0	0	$\frac{1}{2}$	0	0	0	0	0	$\frac{1}{\sqrt{3}}$	$\frac{1}{2\sqrt{3}}$	0
θ_Y^k	0	0	$\frac{-1}{2\sqrt{3}}$	0	$\frac{1}{2}$	0	0	0	0	$\frac{1}{\sqrt{3}}$	$\frac{1}{2\sqrt{3}}$	0	0	0	0	$\frac{-1}{2}$	0	0

FIGURE 6.9
Transformation matrix H (18×18)

6.2 Block Diagonalization of Static Discretized Structure System

6.2.1 Square Plate with Elements $N \times N$

Equally divided into $N \times N$ with equal thickness and isotropic material properties D_4 invariant square plate with even corners supported by point

	v^i_X	v^i_Y	θ^i_Z	v^i_Z	θ^i_X	θ^i_Y	v^j_X	v^j_Y	θ^j_Z	v^j_Z	θ^j_X	θ^j_Y	v^k_X	v^k_Y	θ^k_Z	v^k_Z	θ^k_X	θ^k_Y
v^i_X	$\frac{64D_P}{3}$	0	0	0	0	0	$\frac{-32D_P}{3}$	$\frac{48D_P}{5\sqrt{3}}$	0	0	0	0	$\frac{-32D_P}{3}$	$\frac{-48D_P}{5\sqrt{3}}$	0	0	0	0
v^i_Y	0	$\frac{112D_P}{15}$	0	0	0	0	$\frac{56D_P}{5\sqrt{3}}$	$\frac{-56D_P}{15}$	0	0	0	0	$\frac{-56D_P}{5\sqrt{3}}$	$\frac{-56D_P}{15}$	0	0	0	0
θ^i_Z	0	0	D_T	0	0	0	0	0	$\frac{-D_T}{2}$	0	0	0	0	0	$\frac{-D_T}{2}$	0	0	0
v^i_Z	0	0	0	$\frac{128D_B}{9L^2}$	0	$\frac{-44D_B}{5\sqrt{3}L}$	0	0	0	$\frac{-64D_B}{9L^2}$	$\frac{-32D_B}{9L}$	$\frac{28\sqrt{3}D_B}{45L}$	0	0	0	$\frac{-64D_B}{9L^2}$	$\frac{32D_B}{9L}$	$\frac{28\sqrt{3}D_B}{45L}$
θ^i_X	0	0	0	0	$\frac{208D_B}{135}$	0	0	0	0	$\frac{-122D_B}{45L}$	$\frac{-139D_B}{270}$	$\frac{-59\sqrt{3}D_B}{270}$	0	0	0	$\frac{122D_B}{45L}$	$\frac{-139D_B}{270}$	$\frac{59\sqrt{3}D_B}{270}$
θ^i_Y	0	0	0	$\frac{-44D_B}{5\sqrt{3}L}$	0	$\frac{46D_B}{15}$	0	0	0	$\frac{22D_B}{5\sqrt{3}L}$	$\frac{9D_B}{10\sqrt{3}}$	$\frac{-13D_B}{30}$	0	0	0	$\frac{22D_B}{5\sqrt{3}L}$	$\frac{-9D_B}{10\sqrt{3}}$	$\frac{-13D_B}{30}$
v^j_X	$\frac{-32D_P}{3}$	$\frac{56D_P}{5\sqrt{3}}$	0	0	0	0	$\frac{164D_P}{15}$	$\frac{-52D_P}{5\sqrt{3}}$	0	0	0	0	$\frac{-4D_P}{15}$	$\frac{-4D_P}{5\sqrt{3}}$	0	0	0	0
v^j_Y	$\frac{48D_P}{5\sqrt{3}}$	$\frac{-56D_P}{15}$	0	0	0	0	$\frac{-52D_P}{5\sqrt{3}}$	$\frac{268D_P}{15}$	0	0	0	0	$\frac{4D_P}{5\sqrt{3}}$	$\frac{-212D_P}{15}$	0	0	0	0
θ^j_Z	0	0	$\frac{-D_T}{2}$	0	0	0	0	0	D_T	0	0	0	0	0	$\frac{-D_T}{2}$	0	0	0
v^j_Z	0	0	0	$\frac{-64D_B}{9L^2}$	$\frac{-122D_B}{45L}$	$\frac{22D_B}{5\sqrt{3}L}$	0	0	0	$\frac{128D_B}{9L^2}$	$\frac{22D_B}{5L}$	$\frac{22\sqrt{3}D_B}{15L}$	0	0	0	$\frac{-64D_B}{9L^2}$	$\frac{38D_B}{45L}$	$\frac{-94\sqrt{3}D_B}{45L}$
θ^j_X	0	0	0	$\frac{-32D_B}{9L}$	$\frac{-139D_B}{270}$	$\frac{9D_B}{10\sqrt{3}}$	0	0	0	$\frac{22D_B}{5L}$	$\frac{145D_B}{54}$	$\frac{103\sqrt{3}D_B}{270}$	0	0	0	$\frac{-38D_B}{45L}$	$\frac{-53D_B}{135}$	$\frac{-7\sqrt{3}D_B}{27}$
θ^j_Y	0	0	0	$\frac{28\sqrt{3}D_B}{45L}$	$\frac{-59\sqrt{3}D_B}{270}$	$\frac{-13D_B}{30}$	0	0	0	$\frac{22\sqrt{3}D_B}{15L}$	$\frac{103\sqrt{3}D_B}{270}$	$\frac{173D_B}{90}$	0	0	0	$\frac{-94\sqrt{3}D_B}{45L}$	$\frac{7\sqrt{3}D_B}{27}$	$\frac{-5D_B}{9}$
v^k_X	$\frac{-32D_P}{3}$	$\frac{-56D_P}{5\sqrt{3}}$	0	0	0	0	$\frac{-4D_P}{15}$	$\frac{4D_P}{5\sqrt{3}}$	0	0	0	0	$\frac{164D_P}{15}$	$\frac{52D_P}{5\sqrt{3}}$	0	0	0	0
v^k_Y	$\frac{-48D_P}{5\sqrt{3}}$	$\frac{-56D_P}{15}$	0	0	0	0	$\frac{-4D_P}{5\sqrt{3}}$	$\frac{-212D_P}{15}$	0	0	0	0	$\frac{52D_P}{5\sqrt{3}}$	$\frac{268D_P}{15}$	0	0	0	0
θ^k_Z	0	0	$\frac{-D_T}{2}$	0	0	0	0	0	$\frac{-D_T}{2}$	0	0	0	0	0	D_T	0	0	0
v^k_Z	0	0	0	$\frac{-64D_B}{9L^2}$	$\frac{122D_B}{45L}$	$\frac{22D_B}{5\sqrt{3}L}$	0	0	0	$\frac{-64D_B}{9L^2}$	$\frac{-38D_B}{45L}$	$\frac{-94\sqrt{3}D_B}{45L}$	0	0	0	$\frac{128D_B}{9L^2}$	$\frac{-22D_B}{5L}$	$\frac{22\sqrt{3}D_B}{15L}$
θ^k_X	0	0	0	$\frac{32D_B}{9L}$	$\frac{-139D_B}{270}$	$\frac{-9D_B}{10\sqrt{3}}$	0	0	0	$\frac{38D_B}{45L}$	$\frac{-53D_B}{135}$	$\frac{7\sqrt{3}D_B}{27}$	0	0	0	$\frac{-22D_B}{5L}$	$\frac{145D_B}{54}$	$\frac{-103\sqrt{3}D_B}{270}$
θ^k_Y	0	0	0	$\frac{28\sqrt{3}D_B}{45L}$	$\frac{59\sqrt{3}D_B}{270}$	$\frac{-13D_B}{30}$	0	0	0	$\frac{-94\sqrt{3}D_B}{45L}$	$\frac{-7\sqrt{3}D_B}{27}$	$\frac{-5D_B}{9}$	0	0	0	$\frac{22\sqrt{3}D_B}{15L}$	$\frac{-103\sqrt{3}D_B}{270}$	$\frac{173D_B}{90}$

where $D_P = \frac{E\,t}{1-\nu^2}\,\frac{A}{L^2}$, $D_T = \alpha\,E\,t\,A$, $D_B = \frac{E\,t^3}{12(1-\nu^2)}\,\frac{A}{L^2}$, and $\nu = 0.3$.

FIGURE 6.10
Element stiffness matrix of a regular-trianglar plate K_e (18×18)

	$H^{(1,1)}_{v_{XY}}$	$H^{(1,1)}_{v_Z}$	$H^{(1,1)}_{\theta_{XY}}$	$H^{(1,2)}_{v_{XY}}$	$H^{(1,2)}_{\theta_{XY}}$	$H^{(1,2)}_{\theta_Z}$	$H^{(2,1)}_{v_{XY}}$	$H^{(2,1)}_{v_{XY}}$	$H^{(2,1)}_{v_Z}$	$H^{(2,1)}_{\theta_{XY}}$	$H^{(2,1)}_{\theta_{XY}}$	$H^{(2,1)}_{\theta_Z}$	$H^{(2,1)-}_{v_{XY}}$	$H^{(2,1)-}_{v_{XY}}$	$H^{(2,1)-}_{v_Z}$	$H^{(2,1)-}_{\theta_{XY}}$	$H^{(2,1)-}_{\theta_{XY}}$	$H^{(2,1)-}_{\theta_Z}$
$H^{(1,1)}_{v_{XY}}$	$\frac{208D_P}{5}$	0	0	0	0	0	0	0	0	0	0	0	0	0	0	0	0	0
$H^{(1,1)}_{v_Z}$	0	0	0	0	0	0	0	0	0	0	0	0	0	0	0	0	0	0
$H^{(1,1)}_{\theta_{XY}}$	0	0	$\frac{13D_B}{5}$	0	0	0	0	0	0	0	0	0	0	0	0	0	0	0
$H^{(1,2)}_{v_{XY}}$	0	0	0	0	0	0	0	0	0	0	0	0	0	0	0	0	0	0
$H^{(1,2)}_{\theta_{XY}}$	0	0	0	0	$\frac{7D_B}{5}$	0	0	0	0	0	0	0	0	0	0	0	0	0
$H^{(1,2)}_{\theta_Z}$	0	0	0	0	0	0	0	0	0	0	0	0	0	0	0	0	0	0
$H^{(2,1)}_{v_{XY}}$	0	0	0	0	0	0	0	0	0	0	0	0	0	0	0	0	0	0
$H^{(2,1)}_{v_{XY}}$	0	0	0	0	0	0	0	$\frac{112D_P}{5}$	0	0	0	0	0	0	0	0	0	0
$H^{(2,1)}_{v_Z}$	0	0	0	0	0	0	0	0	$\frac{64D_B}{3L^2}$	$\frac{-76D_B}{15\sqrt{6}L}$	$\frac{64D_B}{3\sqrt{6}L}$	0	0	0	0	0	0	0
$H^{(2,1)}_{\theta_{XY}}$	0	0	0	0	0	0	0	0	$\frac{-76D_B}{15\sqrt{6}L}$	$\frac{61D_B}{45}$	$\frac{-38D_B}{45}$	0	0	0	0	0	0	0
$H^{(2,1)}_{\theta_{XY}}$	0	0	0	0	0	0	0	0	$\frac{64D_B}{3\sqrt{6}L}$	$\frac{-38D_B}{45}$	$\frac{32D_B}{9}$	0	0	0	0	0	0	0
$H^{(2,1)}_{\theta_Z}$	0	0	0	0	0	0	0	0	0	0	0	$\frac{3D_T}{2}$	0	0	0	0	0	0
$H^{(2,1)-}_{v_{XY}}$	0	0	0	0	0	0	0	0	0	0	0	0	0	0	0	0	0	0
$H^{(2,1)-}_{v_{XY}}$	0	0	0	0	0	0	0	0	0	0	0	0	0	$\frac{112D_P}{5}$	0	0	0	0
$H^{(2,1)-}_{v_Z}$	0	0	0	0	0	0	0	0	0	0	0	0	0	0	$\frac{64D_B}{3L^2}$	$\frac{76D_B}{15\sqrt{6}L}$	$\frac{64D_B}{3\sqrt{6}L}$	0
$H^{(2,1)-}_{\theta_{XY}}$	0	0	0	0	0	0	0	0	0	0	0	0	0	0	$\frac{76D_B}{15\sqrt{6}L}$	$\frac{61D_B}{45}$	$\frac{38D_B}{45}$	0
$H^{(2,1)-}_{\theta_{XY}}$	0	0	0	0	0	0	0	0	0	0	0	0	0	0	$\frac{64D_B}{3\sqrt{6}L}$	$\frac{38D_B}{45}$	$\frac{32D_B}{9}$	0
$H^{(2,1)-}_{\theta_Z}$	0	0	0	0	0	0	0	0	0	0	0	0	0	0	0	0	0	$\frac{3D_T}{2}$

FIGURE 6.11
Block-diagonalized element stiffness matrix $\widetilde{K}_e$ (18×18)

bearings. [1] Consider the following; the plate is subjected to asymmetric loading, and the total external force vector of $(H^{\mu})^{\mathrm{T}}\boldsymbol{f}$ is a non-zero element stiffness matrix given by a 4-node Serendipity square element. The stiffness matrix $\widetilde{K}$ transformed by Eq.(5.16) is divided into six blocks.

Fig. 6.12 for the deformation mode $N = 6$ of the solution $H^{\mu}\boldsymbol{w}^{\mu}$ of Eq.(5.20) for each block as seen in the magnitude of the axis indicates the

[1]The finite element solution of this plate is strongly dependent on the mesh partitioning of the small elements.

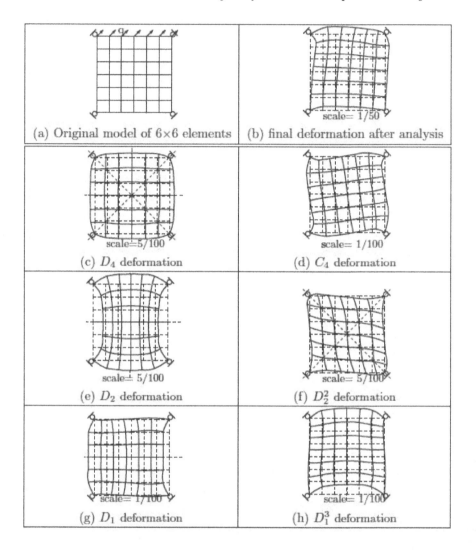

(a) Original model of 6×6 elements | (b) final deformation after analysis

(c) D_4 deformation | (d) C_4 deformation

(e) D_2 deformation | (f) D_2^2 deformation

(g) D_1 deformation | (h) D_1^3 deformation

FIGURE 6.12
Deformation for each symmetry group

magnitude for each mode of deformation. In this case, directly it is obtained from Eq.(3.22). These modes can be superimposed on the solution $\boldsymbol{u}$ of **Fig. 6.12** so the final solution is obtained. In order to argue for the numerical efficiency of the method, a comparison is made between the direct method and this method. A similar comparison was made by Dinkevich [26]. The scavenging cost for a band matrix of size M by the modified Cholesky method,

TABLE 6.1
Number of orbits between elements for a square plate

Parity of N	Type of orbit			
	Center	Square type I	Square type II	Octagonal
odd	1	$(N-1)/2$	$(N-1)/2$	$(N-1)(N-3)/8$
even	0	$N/2$	0	$(N^2-2N)/8$

TABLE 6.2
Number of operations for each element

μ	Type of orbit			
	Center	Square type I	Square type II	Octagonal
$(1,j)_{D_4}$	72	80 or 126	76 or 168	110, 138, 168 or 200
$(2,1)^{\pm}_{D_4}$	152	168 or 300	184 or 400	270, 348, 434 or 528

which is often used for scavenging calculations, is given by the number of

$$B^2(M-B) + B^3/6 \qquad (6.7)$$

operations. where B is the half bandwidth. The cost to the direct method is

$$16N^4 + 127N^3, \qquad (6.8)$$

and the cost of this method can be evaluated at

$$6N^4 + 14N^3. \qquad (6.9)$$

The matrix operation cost for this method can then be calculated. The use of the number of each orbit shown in **Table 6.1** and the cost to the whole board for the cost of each element matrix operation (5.68) shown in **Table 6.2** is obtained by $232N^2$.

This is less than the cost of Eq.(6.8) or (6.9) for the sweep calculation. The total cost of this method is expressed as the sum of Eq.(6.9) and $232N^2$.

$$6N^4 + 14N^3 + 232N^2 \qquad (6.10)$$

This accounts for about $3/8$ of Eq.(6.8) of the cost for a direct method with matrix size N. The numerical efficiency of this method compared to the direct method for the number of partitions N is shown in **Fig. 6.13**. The solid line represents the computation time using the present method without parallel computers, and the dashed line represents the use of the direct method. Computation time is normalized to $N=9$ for the direct method. The present method reduced the computation time for $N=9$ to less than 80% of that taken for the direct method. However, this is not the case for $N=9$, since

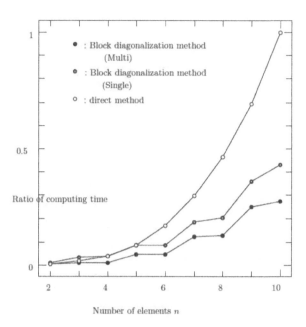

FIGURE 6.13
Computation time for division n

equations (6.8) and (6.10), can be reduced analytically to less than 60%. The reduced matrix creates an additional cost to the block diagonalization analysis in computer program processing. Even taking into account the additional arithmetic cost of (N^2) orders, the total arithmetic cost for an increase in N is reduced by this method compared to the conventional one. Computation time on five computers can be reduced to less than 65%. The computation time is further reduced when the computation for each block is performed by multiple parallel computers. Furthermore, this method can significantly reduce the required array capacity of the computers.

The largest array should be declared for the matrix block $\tilde{K}^\mu$. Furthermore, the array capacity for H^μ can always be discarded by calculating it directly from the H^μ formula. Comparing computational capacities, **Fig. 6.14** requires the required array capacity for the stiffness matrices of both methods, where the direct method represents half the capacity of the zonal matrix. The former, with 30% more storage capacity than the latter, can store a lot of computational capacity.

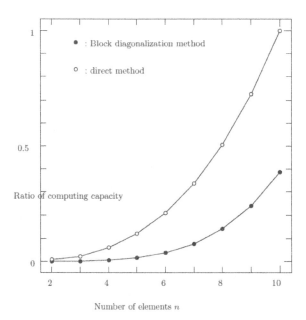

Number of elements n

FIGURE 6.14
Array capacity for split n

6.2.2 D_n Invariant Plate bending Structure Model

Apply the block diagonalization method to the out-of-plane bending linear analysis of a regular n square plate discretized with ACM(Adini,Clough and Melosh) [45] non-conforming (triangular) elements in **Fig. 6.7**. This element is a vertical displacement v_Z^i for each node and a rotation displacement $theta_i$ around the X-axis, it has three degrees of freedom, namely rotational displacement θ_i around the X- and Y-axes, and the shape relation between deflection and deflection angle. The shape functions of deflection and deflection angle are C^1 continuity [2] shall be maintained. As shown in **Fig. 6.15**, the stiffness matrix of a D_n invariant plate (homogeneous in stiffness distribution and material), which is divided into elements without breaking symmetry, is given by the local coordinate transformation matrix consisting of trajectory information regardless of node numbering. It can be transformed into a multiple independent block diagonal matrices. The result is shown in **Fig. 6.16**. Here, [+] in the small block represents a positive value, [·] represents a zero value, and [−] represents a negative value. For example, when $n = 3$, 0 type consists of four trajectories: one at the origin, two 1M type, and one 1V type. When

---

[2]Although the deflection angle between elements does not satisfy continuity, the method of this time is established by the formula (3.23) regardless of whether the elements are relevant or not.

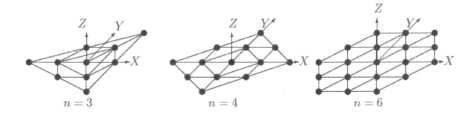

FIGURE 6.15
Discretized D_n invariant thin plate model($n = 3, 4, 6$)

n	$\widetilde{K}^{(1,1)}$	$\widetilde{K}^{(1,2)}$	$\widetilde{K}^{(1,3)}$	$\widetilde{K}^{(1,4)}$	$\widetilde{K}^{(2,1)+} = \widetilde{K}^{(2,1)-}$	$\widetilde{K}^{(2,2)+} = \widetilde{K}^{(2,2)-}$
3	*(matrix)*	*(matrix)*			*(matrix)*	
4	*(matrix)*	*(matrix)*	*(matrix)*	*(matrix)*	*(matrix)*	
6	*(matrix)*	*(matrix)*	*(matrix)*	*(matrix)*	*(matrix)*	*(matrix)*

FIGURE 6.16
Block diagonal matrix $\widetilde{K}$ for rank n.

the order of symmetry is increased to $n = 4, 5, 6, \cdots$, the number of blocks $(m_1 + 2m_2)$ naturally increases accordingly. However, the matrix size per block hardly changes. In addition, each small block also has blocks with orthogonal relationships, but this is accidental and not group-theoretic. **Fig. 6.17** is the total stiffness matrix K for n-gon ($n = 6$), and **Fig. 6.18** is the calculated matrix $\widetilde{K}$ after block diagonalization. Conventionally, the bandwidth of the stiffness matrix increases as n increases, but the bandwidth of the stiffness matrix after block-diagonalization is very small, and the matrix becomes an independent block matrix.

We compared the changes in sequence capacity with increasing n with various methods, and summarized the results in **Fig. 6.19**. The horizontal

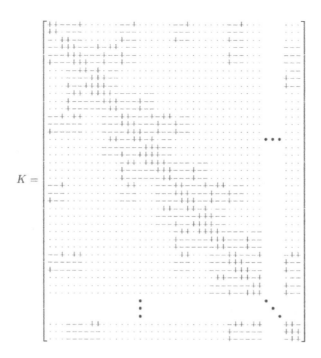

FIGURE 6.17
Global stiffness matrix K

axis is n and the vertical axis is the array capacity of the stiffness matrix K and the ratio of the stiffness matrix $\tilde{K}$ after block diagonalization to the total degrees of freedom. The occupancy rate of the array by block diagonalization (the required array capacity normalized by the square of the total degrees of freedom converges to 0 while oscillating, indicating its advantage). [3] In addition, using EWS-4800, we compared the total computation time associated with the increase of the order n between the conventional method and the new method, **Fig. 6.20** summarizes the results. However, although both cannot simply be measured due to the computer environment such as computer process state, cache RAM, and data input/output time, measurements were made under the same conditions as far as possible. Taking n on the horizontal axis and setting the calculation time to 1 when $n = 20$ by the modified Cholesky method, it can be seen that the present method is more effective than the conventional method as the required computation time from the construction of the stiffness matrix to the coordinate transformation increases by n.

---

[3]n This is because the structure of the irreducible representation differs depending on whether it is even or odd.

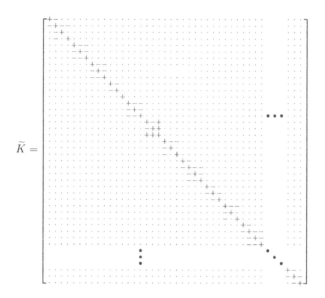

FIGURE 6.18
Global block diagonal matrix $\widetilde{K}$

6.3　Application to Dynamic Structural System

As an example of numerical analysis of this theory, we limit ourselves to the vibration equation as follows

$$M\ddot{u}(t) + C\dot{u}(t) + Ku(t) = f(t) \tag{6.11}$$

of a regular n angular (D_n invariant) structural system, where mass matrix $M = \Gamma_2$, damping matrix $C = \Gamma_1$, stiffness matrix $K = \Gamma_0$, displacement vector $u(t)$, load vector f and t is time (see Eq.(3.3)). The details of the definition of the group D_n and the block diagonalization method of the D_n invariant system are described in the literature [27, 30].

The set of irreducible representations of the group D_6 is

$$R(D_6) = \{(1,1)_{D_6}, (1,2)_{D_6}, (1,3)_{D_6}, (1,4)_{D_6}, (2,1)_{D_6}, (2,2)_{D_6}\} \tag{6.12}$$

where (d, j) represents the j-th d-order irreducible representation. Also, the coordinate transformation matrix corresponding to these irreducible representations has the form

$$H = [H^{(1,1)_{D_6}}, H^{(1,2)_{D_6}}, H^{(1,3)_{D_6}}, H^{(1,4)_{D_6}}, H^{(2,1)_{D_6}}, H^{(2,2)_{D_6}}] \tag{6.13}$$

where the block of the quadratic irreducible representation is divided into two

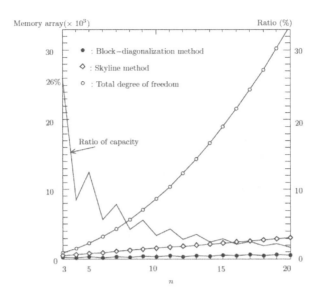

FIGURE 6.19
Relationship between the number of positions n and the required array capacity

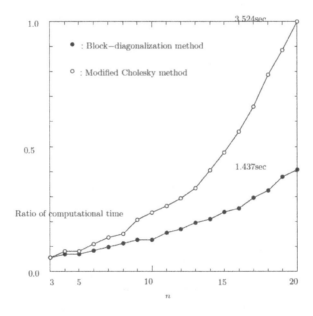

FIGURE 6.20
Relation between the number of places n and the required computation time

blocks shown as

$$H^{(2,j)} = [H^{(2,j)+}, H^{(2,j)-}], \quad j = 1, 2. \tag{6.14}$$

The symmetry of each block matrix of H is shown as

$$\begin{aligned}
\sum H^{(1,1)_{D_6}} &= D_6, & \sum H^{(1,2)_{D_6}} &= C_6 \\
\sum H^{(1,3)_{D_6}} &= D_3, & \sum H^{(1,4)_{D_6}} &= D_3^2 \\
\sum H^{(2,1)_{D_6}^+} &= D_1, & \sum H^{(2,1)_{D_6}^-} &= D_1^4 \\
\sum H^{(2,2)_{D_6}^+} &= D_2, & \sum H^{(2,2)_{D_6}^-} &= C_2
\end{aligned} \tag{6.15}$$

Here, $\sum(\cdot)$ means the group representing the symmetry of the column vectors of the matrix in brackets. A coordinate system corresponding to this coordinate transformation is defined as

$$\begin{aligned}
\boldsymbol{w} = [\quad & (\boldsymbol{w}^{(1,1)_{D_6}})^{\mathrm{T}}, \cdots, (\boldsymbol{w}^{(1,4)_{D_6}})^{\mathrm{T}}, \\
& (\boldsymbol{w}^{(2,1)_{D_6}^+})^{\mathrm{T}}, (\boldsymbol{w}^{(2,1)_{D_6}^-})^{\mathrm{T}}, (\boldsymbol{w}^{(2,2)_{D_6}^+})^{\mathrm{T}}, (\boldsymbol{w}^{(2,2)_{D_6}^-})^{\mathrm{T}}]^{\mathrm{T}}. \tag{6.16}
\end{aligned}$$

For example, the matrix Γ_j of the formula (3.10) is block-diagonalized with

$$\begin{aligned}
H^{\mathrm{T}}\Gamma_j H = \mathrm{diag}[\quad & \widetilde{\Gamma}_j^{(1,1)_{D_6}}, \widetilde{\Gamma}_j^{(1,2)_{D_6}}, \widetilde{\Gamma}_j^{(1,3)_{D_6}}, \widetilde{\Gamma}_j^{(1,4)_{D_6}}, \\
& \widetilde{\Gamma}_j^{(2,1)_{D_6}^+}, \widetilde{\Gamma}_j^{(2,1)_{D_6}^-}, \widetilde{\Gamma}_j^{(2,2)_{D_6}^+}, \widetilde{\Gamma}_j^{(2,2)_{D_6}^-}], \\
& j = 0, 1, 2 \tag{6.17}
\end{aligned}$$

by this coordinate transformation. Two identical blocks correspond to the quadratic irreducible expressions $(2,1)_{D_6}, (2,2)_{D_6}$, respectively.

$$\widetilde{\Gamma}_i^{(2,j)_{D_6}^+} = \widetilde{\Gamma}_i^{(2,j)_{D_6}^-}, \quad i = 0, 1, 2; \quad j = 1, 2 \tag{6.18}$$

Since only one of them needs to be analyzed in the matrix operation, the amount of calculation can be reduced. In addition, matrices B_j and $S(\lambda)$ are also block-diagonalized in the form of Eq.(6.17), so the same argument holds.

6.3.1 D_6 Immutable Beam Structure

The matrices Γ_j $(j = 0, 1, 2)$ of the D_6-invariant consistent mass system model is shown in **Fig. 6.21**(a) and the asymmetric matrix B_j of the D_6-invariant concentrated mass system model is shown in **Fig. 6.21**(b) and block diagonalized by the present method. This system has a total of 7 degrees of freedom with 1 node and 1 degree of freedom for displacement in the z direction. In this case, since there are not enough degrees of freedom, the space of the first-order irreducible representations $(1,2)_{D_6}, (1,4)_{D_6}$ of the group D_6 is degenerated, and the remaining four become irreducible. There are 6 block matrices corresponding to the expression $(1,1)_{D_6}, (1,3)_{D_6}, (2,1)_{D_6}, (2,2)_{D_6}$.

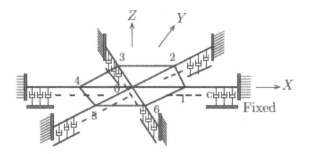

FIGURE 6.21

D_6 immutable structural model

6.3.2 Simultaneous Block Diagonalization of Various Matrices

The matrices K, C, and M of the D_6 invariant structural model in **Fig. 6.21**(a) are shown below.

$$
K = K_0 \begin{pmatrix}
6 & -1 & -1 & -1 & -1 & -1 & -1 \\
-1 & 4 & -1 & & & & -1 \\
-1 & -1 & 4 & -1 & & & \\
-1 & & -1 & 4 & -1 & & \\
-1 & & & -1 & 4 & -1 & \\
-1 & & & & -1 & 4 & -1 \\
-1 & -1 & & & & -1 & 4
\end{pmatrix}
\tag{6.19}
$$

$$
C = C_0 \begin{pmatrix}
6 & -1 & -1 & -1 & -1 & -1 & -1 \\
-1 & 2 & & & & & \\
-1 & & 2 & & & & \\
-1 & & & 2 & & & \\
-1 & & & & 2 & & \\
-1 & & & & & 2 & \\
-1 & & & & & & 2
\end{pmatrix}
\tag{6.20}
$$

$$
M = M_0 \begin{pmatrix}
12 & -1 & -1 & -1 & -1 & -1 & -1 \\
-1 & 8 & -1 & & & & -1 \\
-1 & -1 & 8 & -1 & & & \\
-1 & & -1 & 8 & -1 & & \\
-1 & & & -1 & 8 & -1 & \\
-1 & & & & -1 & 8 & -1 \\
-1 & -1 & & & & -1 & 8
\end{pmatrix}
\tag{6.21}
$$

where, $K_0 = 12EI/\ell^3, C_0 = \nu A\ell, M_0 = \rho A\ell/6$. This damping matrix C is neither proportional damping nor Rayleigh damping, so it is difficult to handle in general modal analysis. These matrices are block-diagonalized simultaneously with

$$\widetilde{K} = H^{\mathrm{T}} K H$$

$$= K_0 \begin{pmatrix} \begin{array}{|cc|} \hline 6 & -\sqrt{6} \\ -\sqrt{6} & 2 \\ \hline \end{array} & & & & & O \\ & \boxed{6} & & & & \\ & & \boxed{3} & & & \\ & & & \boxed{3} & & \\ & & & & \boxed{5} & \\ & O & & & & \boxed{5} \\ \end{pmatrix} \qquad (6.22)$$

$$\widetilde{C} = H^{\mathrm{T}} C H$$

$$= C_0 \begin{pmatrix} \begin{array}{|cc|} \hline 6 & -\sqrt{6} \\ -\sqrt{6} & 2 \\ \hline \end{array} & & & & & O \\ & \boxed{2} & & & & \\ & & \boxed{2} & & & \\ & & & \boxed{2} & & \\ & & & & \boxed{2} & \\ & O & & & & \boxed{2} \\ \end{pmatrix} \qquad (6.23)$$

$$\widetilde{M} = H^{\mathrm{T}} M H$$

$$= M_0 \begin{pmatrix} \begin{array}{|cc|} \hline 12 & -\sqrt{6} \\ -\sqrt{6} & 6 \\ \hline \end{array} & & & & & O \\ & \boxed{10} & & & & \\ & & \boxed{7} & & & \\ & & & \boxed{7} & & \\ & & & & \boxed{9} & \\ & O & & & & \boxed{9} \\ \end{pmatrix} \qquad (6.24)$$

by the coordinate transformation matrix H shown in **Table 6.3**.

Also, the differential Eq.(3.2) can be decomposed into an expression for each

$$M^\mu \ddot{w}^\mu + C^\mu \dot{w}^\mu + K^\mu w^\mu = (H^\mu)^{\mathrm{T}} f,$$
$$\mu = (1,1)_{D_6}, \ (1,3)_{D_6}, \ (2,1)^{\pm}_{D_6}, \ (2,2)^{\pm}_{D_6} \qquad (6.25)$$

and an irreducible expression (the expression corresponding to the second-order irreducible expression is divided into two).

6.3.3 Vibration Analysis of Concentrated Mass System without Attenuation

The complex natural frequency λ of the structural model in **Fig. 6.21** is given as the solution of $|S(\lambda)| = 0$. Block-diagonalizing $S(\lambda)$ of this structure using

TABLE 6.3
D_6 invariant coordinate transformation matrix H^μ (z direction component)

No.	$(1,1)_{D_6}$	$(1,3)_{D_6}$	$(2,1)^+_{D_6}$	$(2,1)^-_{D_6}$	$(2,2)^+_{D_6}$	$(2,2)^-_{D_6}$
0	1	0	0	0	0	0
1	0	$1/\sqrt{6}$	$1/\sqrt{6}$	$1/\sqrt{3}$	0	$1/\sqrt{3}$	0
2	0	$1/\sqrt{6}$	$-1/\sqrt{6}$	$\sqrt{3}/6$	$1/2$	$-\sqrt{3}/6$	$1/2$
3	0	$1/\sqrt{6}$	$1/\sqrt{6}$	$-\sqrt{3}/6$	$1/2$	$-\sqrt{3}/6$	$-1/2$
4	0	$1/\sqrt{6}$	$-1/\sqrt{6}$	$-1/\sqrt{3}$	0	$1/\sqrt{3}$	0
5	0	$1/\sqrt{6}$	$1/\sqrt{6}$	$-\sqrt{3}/6$	$-1/2$	$-\sqrt{3}/6$	$1/2$
6	0	$1/\sqrt{6}$	$-1/\sqrt{6}$	$\sqrt{3}/6$	$-1/2$	$-\sqrt{3}/6$	$-1/2$

the coordinate transformation matrix H of **Table** 6.3 yields

$$\widetilde{S}(\lambda) = \mathrm{diag}[\widetilde{S}^{(1,1)_{D_6}}, \widetilde{S}^{(1,3)_{D_6}}, \widetilde{S}^{(2,1)_{D_6}}, \widetilde{S}^{(2,2)_{D_6}}] \qquad (6.26)$$

where,

$$\widetilde{S}^{(2,j)_{D_6}} = \mathrm{diag}[\widetilde{S}^{(2,j)^+_{D_6}}, \widetilde{S}^{(2,j)^-_{D_6}}], \qquad (6.27)$$

$$\widetilde{S}^{(2,j)^+_{D_6}} = \widetilde{S}^{(2,j)^-_{D_6}}, \quad j = 1, 2 \qquad (6.28)$$

There are two identical detail blocks in the quadratic irreducible representation. Thus, the problem of finding the solution λ of the function $|S(\lambda)|$ can be rewritten as the problem of finding the solution of four independent equations as follows;

$$|\widetilde{S}^\mu(\lambda)| = 0, \quad \mu = (1,1)_{D_6}, (1,2)_{D_6}, (2,1)^+_{D_6}, (2,2)^+_{D_6}. \qquad (6.29)$$

That is, the multiple root of $|S(\lambda)| = 0$ is $|\widetilde{S}^{(2,1)^+_{D_6}}| = 0$ or $|\widetilde{S}^{(2,2)^+_{D_6}}| = 0$.

Now, if M is a unit matrix and non-decaying, then λ is a real root. Plot $|S(\lambda)|$ and $|\widetilde{S}^\mu(\lambda)|$ as functions of λ in **Fig. 6.22**(a) and (b) respectively. The solution λ of $|S(\lambda)| = 0$ corresponds to any $|\widetilde{S}^\mu(\lambda)| = 0$ solution and it is clear that they give the same eigenvalue λ. $|S(\lambda)|$ is a more complex function than $|\widetilde{S}^\mu(\lambda)|$, which increases and decreases more rapidly, and is tangent to the abscissa in terms of the eigenvalues of the heavy roots. These properties are destabilising factors for iterative calculations. On the other hand, $|\widetilde{S}^\mu(\lambda)|$ is a relatively simple and smooth function and the eigenvalues of the multiple roots λ are distributed in different blocks, so $|\widetilde{S}^\mu(\lambda)|$ for each block always intersects the horizontal axis. Thus, the block diagonalization method has the effect of stabilising the numerical analysis by dispersing the singularity of $|S(\lambda)| = 0$. It is also highly advantageous for systems with higher-order symmetries, as it increases the number of blocks that are identical.

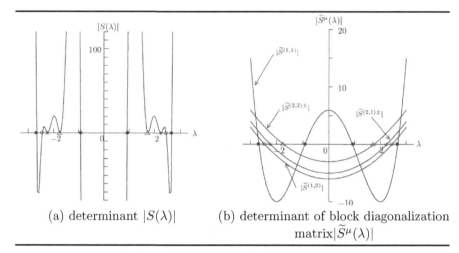

(a) determinant $|S(\lambda)|$ 　　　(b) determinant of block diagonalization
　　　　　　　　　　　　　　　　matrix$|\widetilde{S}^{\mu}(\lambda)|$

FIGURE 6.22

D_6 Comparison of determinants of invariant lumped-mass systems $|S(\lambda)|$ and $|\widetilde{S}^{\mu}(\lambda)|$

6.3.4　Vibration Analysis of Concentrated Mass system with Damping

We performed a complex eigenvalue analysis of the complex dynamic stiffness matrix $S(\lambda)$ of the model shown in **Fig. 6.21**(b) (Eq.(6.34) and (6.20)) $K_0 = 1(\text{N/m})$, $C_0 = 0.1(\text{Ns/m})$ and $M = I(\text{kg})$). The absolute value of the determinant $|S(\lambda)|$ for the real part $\text{Re}(\lambda)$ and the imaginary part $\text{Im}(\lambda)$ of λ three-dimensional plot is **Fig. 6.23**(a). The solution of $|S(\lambda)| = 0$ is marked with ($\bullet$) in the figure. In addition, the symbols s1 and s3 in the figure represent first-order irreducible representations, and d1 and d2 represent solutions corresponding to second-order irreducible representations. The solutions corresponding to quadratic irreducible representations are repeated roots. **Fig. 6.23**(b) shows contours of the absolute value of $|S|$ in the $\text{Im}(\lambda) < 0$ region. Clearly, $|S(\lambda)|$ is a very complicated function.

On the other hand, the absolute value of the complex dynamic stiffness matrix $|\widetilde{S}^{\mu}(\lambda)|$ of the formula (6.25) decomposed for each block is **Fig. 6.24**(a)-(d). Compared to **Fig. 6.23**, they are all decomposed into very smooth functions, which naturally agree with existing solutions. The solution of the repeated root is $|\widetilde{S}^{(2,1)\pm}| = 0$ or $|\widetilde{S}^{(2,2)\pm}| = 0$ (see **Fig. 6.24**(c),(d)).

As an example of the comparison of eigenvalue analysis of $S(\lambda)$ and $\widetilde{S}^{\mu}(\lambda)$, **Fig. 6.32** shows the convergence of the Newton-Raphson solution for various initial values. Although the convergence of the solution strongly depends on the initial value, the convergence of the solution of this method is generally better than that of the conventional method.

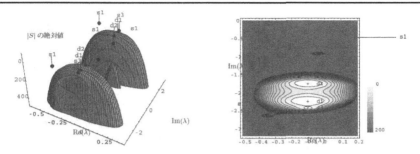

(a) 3D diagram of the absolute value of $|S(\lambda)|$(b) Contour plot of absolute value of $|S(\lambda)|$

FIGURE 6.23
Distribution of the absolute value of the determinant $|S(\lambda)|$ for the complex eigenvalue λ

6.4 Applying to D_n Invariant Linear Vibration System

As an example of numerical analysis, we will consider a second-order ordinary differential equation for a decaying linear oscillation system under periodic external forces that is invariant to the Dihedral group D_4:

$$M\ddot{\boldsymbol{x}} + C\dot{\boldsymbol{x}} + K\boldsymbol{x} = \boldsymbol{f}\cos\omega t \tag{6.30}$$

The details about the definition of the group D_n and the block diagonalization method for D_n-invariant systems will be left to the references [27, 30] and **Section 6.3**.

6.4.1 Block Diagonalization Example of D_4 Invariant Frame Structure

As a simple structural example that demonstrates the outline of this method, consider the three-dimensional frame structure invariant to D_4 shown in **Fig. 6.25**. This system has a total of 8 degrees of freedom, consisting only of horizontal (XY) direction displacements of 1 node 2 degrees of freedom. The bending stiffness k_x, k_y and damping c_x, c_y of the columns are both considered to be constant. The set of irreducible representations of the group D_4 is, from **Section 6.3**,

$$R(D_4) = \{(1,1)_{D_4}, (1,2)_{D_4}, (1,3)_{D_4}, (1,4)_{D_4}, (2,1)_{D_4}\} \tag{6.31}$$

Here, $(d,j)_{D_n}$ represents the j-th d-dimensional irreducible representation for the group D_n.

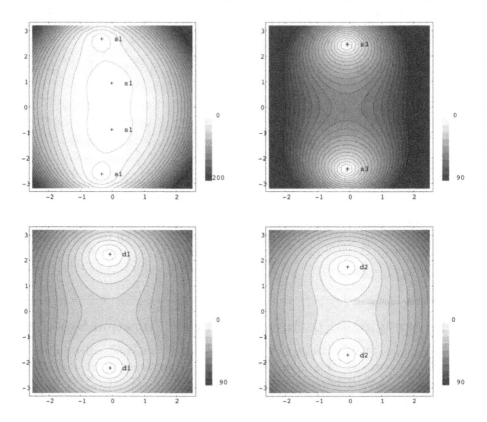

FIGURE 6.24

Determinant $|S^{\mu}(\lambda)|$ for complex eigenvalue λ after block diagonalization

The linear decaying oscillation equation under periodic external forces for the D_4 invariant structural system in **Fig. 6.25**. Eq.(6.30) can be represented as follows.

$$\left\{ \begin{array}{c} \dot{x} \\ \ddot{x} \end{array} \right\} = \left[\begin{array}{cc} O & I \\ \Gamma_0 & \Gamma_1 \end{array} \right] \left\{ \begin{array}{c} x \\ \dot{x} \end{array} \right\} + \left[\begin{array}{c} O \\ -M^{-1} \end{array} \right] f \cos \omega t \qquad (6.32)$$

In other words, it can be expressed in the form of a state equation as

$$\frac{\mathrm{d}\widehat{x}}{\mathrm{d}t} = A\widehat{x} + \overline{B}u \qquad (6.33)$$

Here, $\Gamma_0 = -M^{-1}K$, $\Gamma_1 = -M^{-1}C$, $u(t) = f \cos \omega t$ and let's set $\gamma = \overline{B}f$.

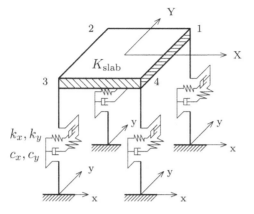

FIGURE 6.25
A structural model invariant to D_4

Now, it is assumed that each matrix and/or varible Γ_j shows

$$
\Gamma_0 = \alpha_1
\begin{pmatrix}
54 & 13 & -40 & -1 & 0 & -13 & -14 & 1 \\
 & 54 & 1 & -14 & -13 & 0 & -1 & -40 \\
 & & 54 & -13 & -14 & -1 & 0 & 13 \\
 & & & 54 & 1 & -40 & 13 & 0 \\
 & & & & 54 & 13 & -40 & -1 \\
 & & & & & 54 & 1 & -14 \\
 & \text{Symm.} & & & & & 54 & -13 \\
 & & & & & & & 54
\end{pmatrix}
$$

$$
+\alpha_2
\begin{pmatrix}
4 & 0 & -2 & 0 & 1 & 0 & -2 & 0 \\
 & 4 & 0 & -2 & 0 & 1 & 0 & -2 \\
 & & 4 & 0 & -2 & 0 & 1 & 0 \\
 & & & 4 & 0 & -2 & 0 & 1 \\
 & & & & 4 & 0 & -2 & 0 \\
 & & & & & 4 & 0 & -2 \\
 & \text{Symm.} & & & & & 4 & 0 \\
 & & & & & & & 4
\end{pmatrix}
$$

$$
\Gamma_1 = \alpha_3
\begin{pmatrix}
8 & 0 & -2 & 0 & 1 & 0 & -2 & 0 \\
 & 8 & 0 & -2 & 0 & 1 & 0 & -2 \\
 & & 8 & 0 & -2 & 0 & 1 & 0 \\
 & & & 8 & 0 & -2 & 0 & 1 \\
 & & & & 8 & 0 & -2 & 0 \\
 & & & & & 8 & 0 & -2 \\
 & \text{Symm.} & & & & & 8 & 0 \\
 & & & & & & & 8
\end{pmatrix}
$$

here,

$$
\alpha_1 = -\frac{15Et_{\text{slab}}}{91M_0}, \quad \alpha_2 = -\frac{364p_0Et_{\text{slab}}}{91M_0}, \quad \alpha_3 = -\frac{4C_0}{M_0}, \quad p_0 = \frac{EI_{\text{col}}/\ell^3}{Et_{\text{slab}}}
$$

$$
\Gamma_0 = -(M_0M_{\text{slab}})^{-1}(K_{\text{slab}} + K_0K_{\text{col}}), \quad \Gamma_1 = -(M_0M_{\text{slab}})^{-1}(C_0C_{\text{col}})
$$

$K_0 = 12EI/\ell^3$, $C_0 = \nu A\ell$, $M_0 = \rho V/36$ $K_{\text{col}}, C_{\text{col}}, M_{\text{slab}}$ are shown the following:

$$K_{\text{col}} = C_{\text{col}} = \text{diag}[1,1,1,1,1,1,1,1],$$

$$M_{\text{slab}} = \begin{pmatrix} 4 & 0 & 2 & 0 & 1 & 0 & 2 & 0 \\ & 4 & 0 & 2 & 0 & 1 & 0 & 2 \\ & & 4 & 0 & 2 & 0 & 1 & 0 \\ & & & 4 & 0 & 2 & 0 & 1 \\ & & & & 4 & 0 & 2 & 0 \\ & & & & & 4 & 0 & 2 \\ & & \text{Symm.} & & & & 4 & 0 \\ & & & & & & & 4 \end{pmatrix}$$

The displacement vector is

$$x(t) = \{x_1,\ y_1,\ \cdots,\ x_4,\ y_4\}^{\text{T}} \tag{6.34}$$

and is assumed to correspond to each node. By performing a coordinate transformation according to Eq.(6.33) becomes as follows.

$$\left\{ \begin{array}{c} \dot{w} \\ \ddot{w} \end{array} \right\} = \tilde{A} \left\{ \begin{array}{c} w \\ \dot{w} \end{array} \right\} + H_X^{\text{T}} \gamma \cos \omega t \tag{6.35}$$

At this time, $\tilde{A}$ becomes as follows.

$$\tilde{A} = \left(\begin{array}{c|c} O & I \\ \hline \text{diag}[\cdots, \Gamma_0^\mu, \cdots] & \text{diag}[\cdots, \Gamma_1^\mu, \cdots] \end{array} \right), \quad \mu \in R(D_4) \tag{6.36}$$

It decomposes into each irreducible representation. Therefore, the linear equation of the D_4 invariant structural system becomes as follows.

$$\left\{ \begin{array}{c} \dot{w}^\mu \\ \ddot{w}^\mu \end{array} \right\} = A^\mu \left\{ \begin{array}{c} w^\mu \\ \dot{w}^\mu \end{array} \right\} + (H_X^\mu)^{\text{T}} \gamma \cos \omega t,$$

$$\mu = (1,1)_{D_4}, (1,2)_{D_4}, (1,3)_{D_4}, (1,4)_{D_4}, (2,1)_{D_4}^+, (2,1)_{D_4}^- \tag{6.37}$$

It decomposes into independent differential equations. In this decomposition, $\tilde{\Gamma}_j$ ($j = 0, 1$) of this system is given by Eq.(5.35).

$$\tilde{\Gamma}_0 = -(H_X^{\text{T}} M^{-1} H_X)(H_X^{\text{T}} K H_X)$$

$$= 4\alpha_1 \begin{pmatrix} 26 & & & & & & & \\ & 0 & & & & & & \\ & & 14 & & & & O & \\ & & & 14 & & & & \\ & & & & 0 & 0 & & \\ & & & & 0 & 27 & & \\ & & O & & & & 0 & 0 \\ & & & & & & 0 & 27 \end{pmatrix} + \alpha_2 \begin{pmatrix} 3 & & & & & & & \\ & 3 & & & & & & \\ & & 3 & & & & O & \\ & & & 3 & & & & \\ & & & & 1 & 0 & & \\ & & & & 0 & 9 & & \\ & & O & & & & 1 & 0 \\ & & & & & & 0 & 9 \end{pmatrix}$$

$$\tag{6.38}$$

$$\widetilde{\Gamma}_1 = -(H_X^{\mathrm{T}} M^{-1} H_X)(H_X^{\mathrm{T}} C H_X)$$

$$= \alpha_3 \begin{pmatrix} \boxed{7} & & & & & & \\ & \boxed{7} & & & O & & \\ & & \boxed{7} & & & & \\ & & & \boxed{7} & & & \\ & & & & \boxed{\begin{matrix}5 & 0\\0 & 13\end{matrix}} & & \\ & O & & & & \boxed{\begin{matrix}5 & 0\\0 & 13\end{matrix}} \end{pmatrix} \qquad (6.39)$$

It is transformed into a block diagonal form. The deformation modes corresponding to each irreducible representation are shown in **Fig. 6.26**.

For example, let's set the initial values of the differential equation (6.33) to

$$x(0) = \mathbf{0}, \quad \dot{x}(0) = \mathbf{0} \qquad (6.40)$$

and let's also set the asymmetric external force vector to

$$\gamma = \{\, 0,\ 0,\ 0,\ 1,\ 0,\ 0,\ 0,\ 1 \,\}^{\mathrm{T}} \qquad (6.41)$$

In this case, the external force vector on the right side of Eq.(6.37), $(H_X^{\mu})^{\mathrm{T}} \gamma$, becomes zero for all irreducible representations except for $(1,1)_{D_4}$ and $(1,4)_{D_4}$. The results of solving Eq.(6.37), decomposed for each irreducible representation, using the Runge-Kutta method are shown in **Fig. 6.27**. **Fig. 6.27**(a) and (b) each show the oscillation waveforms of the irreducible representations $\mu = (1,1)_{D_4}$ and $(1,4)_{D_4}$, respectively. In contrast, **Fig. 6.27**(c) is the result of directly numerically solving the system of differential equations in Eq.(6.33) using the Runge-Kutta method. Naturally, the solution is identical to the one obtained by superposing the solutions for $\mu = (1,1)_{D_4}$ and $(1,4)_{D_4}$ in our method.

The calculation time for the Runge-Kutta method was, on average, 1.18 sec/block for our method, and 17.00 sec for the direct method. Thus, the solution could be found in less than 7% of the calculation time compared to conventional methods (calculations were done using the EWS4800/310 workstation).

6.4.2 D_n Invariant m-Layer Dome Structure

We apply our method to the m-layer D_n invariant symmetric structure shown in **Fig. 6.28**. We take the intersection points of the members of this dome as nodes, and fix them at the contact points ($Z = 0$). The stiffness EA and the unit volume weight w are the same for all members, and the mass is

$$m_{ij} = \frac{wA\ell}{6} \begin{pmatrix} 2 & 1 \\ 1 & 2 \end{pmatrix}$$

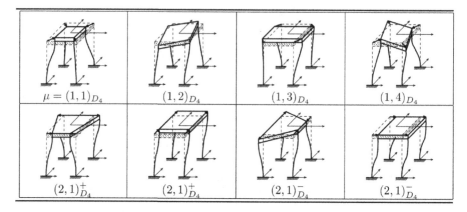

FIGURE 6.26

Invariant oscillation modes of each irreducible representation of D_4 (XY direction components)

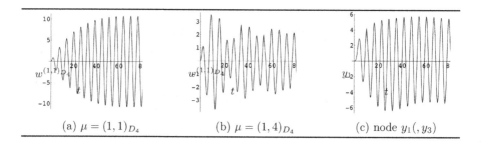

FIGURE 6.27

Phase orbits of displacement and velocity for each irreducible representation of D_4-invariant structural system ($\omega =$ 1.0rad/sec)

using the equivalent mass. The damping is considered as a distributed damping system model, uniformly applying 0.3(Ns/m^2) per unit length of the member, and 0.3(Ns/m) on each node.

The total degrees of freedom for this system is

$$N = 2mn \qquad (6.42)$$

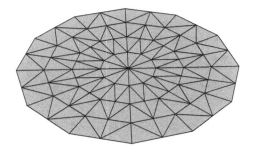

FIGURE 6.28
Regular n polygonal m layered structure ($n = 12, m = 4$)

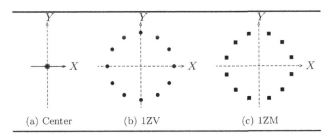

(a) Center   (b) 1ZV   (c) 1ZM

FIGURE 6.29
Orbit decomposition diagram ($n = 12$)

and the number of blocks $q(\mathcal{G})$ for the group D_n can be approximated by

$$q(D_n) = n + \frac{(-1)^n - 1}{2} \simeq n. \tag{6.43}$$

4

By the way, classifying the nodes of this structural system into orbits (see **Fig. 6.29** [27, 30]), there is one Center type at the zenith, and 1ZV type and 1ZM type consist of m and $m - 1$ respectively, a total of $2m$. The matrix size N^μ for a given irreducible representation μ is greater than the number of orbits, thus,

$$N^\mu = \begin{cases} 2m & \text{for} \quad \mu = (1,1) \\ 2m - 1 & \text{for} \quad \mu \neq (1,1) \end{cases}$$
$$\simeq 2m \tag{6.44}$$

From equations (6.59) to (6.44), it can be seen that the approximation of

---

[4]The reason why the number of blocks is less than that from Eq.(5.16) is because the number of blocks is reduced due to degeneracy in space.

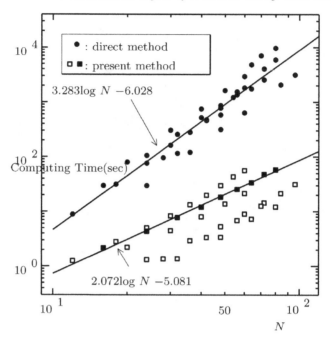

FIGURE 6.30
Comparison of computation times for the Runge-Kutta method with respect
to N

Eq.(7.6) holds. If Eq.(6.60) is substituted into Eq.(7.8), the ratio of the computational complexity of the direct method to our method is

$$\frac{\tilde{\rho}}{\rho} \simeq \frac{1}{n^{\alpha-1}}. \tag{6.45}$$

For example, if n and m are varied to satisfy

$$n = c\,m \tag{6.46}$$

where $c > 0$, then Eq.(6.59) becomes $N = (2/c)n^2$, and Eq.(6.45) becomes

$$\frac{\tilde{\rho}}{\rho} = \left(\frac{2}{cN}\right)^{\frac{\alpha-1}{2}}. \tag{6.47}$$

Using a NEC workstation EWS–4800, we conducted a vibration analysis using the Runge-Kutta method when a periodic external force of $\cos t$ $(0 \le t \le 20\pi$, $dt = 0.002\text{sec})$ was applied in the vertical direction to the zenith node. The computation times for the direct method and our method when the parameters n, m of the D_n invariant m-level dome structure are

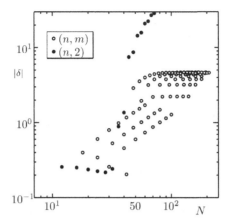

FIGURE 6.31
Relative error ratio with change in N ($2\leq n \leq 20$, $2\leq m \leq 10$)

varied are shown in **Fig. 6.30**. In the figure, the ($\bullet$) symbol represents the computation time by the traditional direct method, while ($\square$) and ($\blacksquare$) represent the computation time by our method. In particular, ($\blacksquare$) represents data satisfying $n = 2m$ (where $c = 2$ in Eq.(6.62)). The straight lines in the figure represent regression lines, and the regression line for our method is calculated for data satisfying $n = 2m$. The computation time is roughly proportional to $N^3(\alpha = 3)$ for the direct method, and to N^2 for our method when $c = 2$, so the ratio of computation efficiencies is $\tilde{\rho}/\rho \propto 1/N$, which is consistent with the result of Eq.(6.63) when $\alpha = 3$.

These results corroborate the validity of the evaluation formula (6.63) for the comparison of computational efficiency between the conventional method and our method, which was derived based on assumption (6.62). The other data ($\square$) in **Fig. 6.30** also approximately follow the regression line, suggesting that the above discussion can be applied as is.

The relative error ratio $|\delta|$ in the eigenvalue analysis for both methods was examined as the geometric parameters (n, m) of this dome structure were varied, and the matrix size N correspondingly changed. In cases where the eigenvalue is separated from other eigenvalues, it depends on the floating point processing ability of the computer, but the eigenvalue almost satisfies machine precision (16 digits), and no difference was observed between the two methods.

However, the case of eigenvalues that become multiple roots was examined. The change in the error ratio of the two methods as N increases is shown in **Fig. 6.31**. The points marked with ($\circ$) in the figure represent the relative error ratio for $(n, m)(m \geq 3)$, and the points marked with ($\bullet$) represent specifically the points for $(n, 2)$.

With the increase in N, the relative error ratio $|\delta|$ increases sharply, demonstrating the superiority of our method. This is consistent with the backward error analysis. There is a tendency for the relative error to increase with the increase of the parameters (n, m). Furthermore, the error tended to increase more due to the increase in n rather than the increase in m. This is due to the fact that the increase in n increases the number of blocks and causes a reduction in block size.

6.4.2.1 Eigenvalue Error Estimation by Backward Error Analysis

Consider the transformation of matrix A into a Hessenberg matrix. The norm of the rounding error δA in this case, is represented by backward error analysis as

$$||\delta A|| \leq 2^{-q} a N^2 ||A||. \tag{6.48}$$

Here, a represents a constant of order 1 and q represents the number of binary floating point digits. Investigate the impact of the rounding error δA of matrix A on its eigenvalues. Let α_i be a single eigenvalue of A and $\boldsymbol{\xi}_i$ and $\boldsymbol{\eta}_i$ be the corresponding normalized right and left eigenvectors, respectively ($||\boldsymbol{\xi}_i|| = ||\boldsymbol{\eta}_i|| = 1$). When the noise δA approaches the zero matrix, $A + \delta A$ has a single eigenvalue $\alpha_i + \delta\alpha_i$ such that

$$\alpha_i + \delta\alpha_i \approx \frac{\boldsymbol{\eta}_i^T (A + \delta A)\boldsymbol{\xi}_i}{\boldsymbol{\eta}_i^T \boldsymbol{\xi}_i}. \tag{6.49}$$

For sufficiently small δA, for example,

$$|\delta\alpha_i| \leq \frac{||\delta A||}{|\boldsymbol{\eta}_i^T \boldsymbol{\xi}_i|}$$

holds. Using Eq.(6.48), the above formula becomes

$$|\delta\alpha_i| \leq \frac{2^{-q} a N^2 ||A||}{|\boldsymbol{\eta}_i^T \boldsymbol{\xi}_i|} \tag{6.50}$$

Here, $\cos\theta_i$ is the direction cosine between $\boldsymbol{\eta}_i$ and $\boldsymbol{\xi}_i$. For eigenvalues after the decomposition into irreducible representations, it is represented as

$$|\delta\alpha_i^\mu| \leq \frac{2^{-q} a (N^\mu)^2 \sqrt{\sum \sum_{i,j=1}^{N^\mu} (a_{ij}^\mu)^2}}{|\cos\theta_i^\mu|} \tag{6.51}$$

Taking the ratio of the upper limits of relative errors using Eqs.(6.50) and (6.51),

$$\begin{aligned} \delta &= \left| \frac{(\delta\alpha_i^\mu)_{\max}}{(\delta\alpha_i)_{\max}} \right| \\ &= \left| \frac{(N^\mu)^2 \sqrt{\sum \sum_{i,j=1}^{N^\mu} (a_{ij}^\mu)^2}}{N^2 \sqrt{\sum \sum_{i,j=1}^{N} a_{ij}^2}} \frac{\cos\theta_i}{\cos\theta_i^\mu} \right| \end{aligned} \tag{6.52}$$

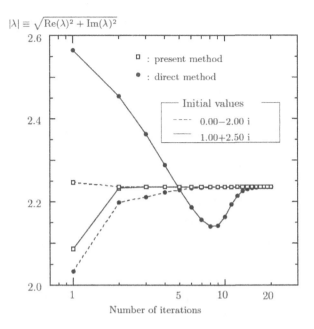

FIGURE 6.32
Comparison of convergence performance by Newton-Raphson method

To simplify this evaluation formula, assume that the components of matrices A, A^μ are simply Δ, $\cos\theta_i / \cos\theta_i^\mu \approx 1$, using Eq.(7.6), Eq.(6.52) becomes

$$\delta \cong \left| \frac{(N^\mu)^3 \Delta}{N^3 \Delta} \right| = \frac{1}{[q(D_n)]^3} \tag{6.53}$$

Therefore, by using this method, the relative error can be reduced to about $1/[q(D_n)]^3$.

6.4.3 D_n Invariant m-layer Structure Eigenvalue Analysis

This method is applied to the eigenvalues problem for the control matrix A of D_n invariant positive n polygonal m layer structure (1 node, 1 vertical degree of freedom) shown in **Fig. 6.28**. It is placed at an attenuation value of 0.1 (Ns/m) on the element on the ray. This is neither proportional damping nor Rayleigh damping.

6.4.4 Analytical Evaluation of Computational Efficiency

Let us analytically examine the efficiency of the eigenvalue analysis of this model. The main term of the total complexity is denoted

$$p = aN^3 \tag{6.54}$$

for matrix size N. where a is a constant. On the other hand, the main term of the computational complexity in the block diagonalization method is

$$\widetilde{p} = \sum_{\mu \in R(G)} a(N^\mu)^3 \tag{6.55}$$

where N^μ represents the matrix size of each irreducible representation space. The amount of computation associated with coordinate transformation of Eq.(5.35). Additionally, suppose that all blocks K^μ have the same size N^μ, and approximate N^μ with

$$N^\mu \simeq \frac{N}{q(G)}. \tag{6.56}$$

where $q(G)$ represents the total number of blocks in the group G. Also, by substituting the expression (7.6) into the expression (7.5),

$$\widetilde{p} \simeq a\left(\frac{N^3}{[q(G)]^2}\right) \tag{6.57}$$

can be approximated. From the formulas (7.3) and (7.7), the ratio of computational costs between the proposed method and the conventional method is

$$\frac{\widetilde{p}}{p} \simeq \frac{1}{[q(G)]^2} \tag{6.58}$$

approximately. However, since N is also a function of $q(G)$, detailed examination of each structure is necessary.

For example, the total degrees of freedom N of this structural system is denoted by

$$N = 2mn + 1 \simeq 2mn \tag{6.59}$$

The number of blocks $q(G)$ in the group D_n is

$$q(D_n) = n + \frac{(-1)^n - 1}{2} \simeq n \tag{6.60}$$

(this formula takes into account the reduction in the number of blocks due to spatial degeneration). Substituting Eq.(6.60) into Eq.(7.7) and Eq.(7.8), it becomes

$$\widetilde{p} = a\frac{N^3}{n^2}, \qquad \frac{\widetilde{p}}{p} \simeq \frac{1}{n^2} \tag{6.61}$$

If n is large enough, our method can reduce the computational cost to about $1/n^2$.

Also, consider a change of n and m to satisfy the following.

$$m = cn \tag{6.62}$$

At this time, Eq.(6.59) becomes $N = 2cn^2$ and Eq.(6.61) becomes

$$\widetilde{p} = 2acN^2, \qquad \frac{\widetilde{p}}{p} = \frac{2c}{N} \tag{6.63}$$

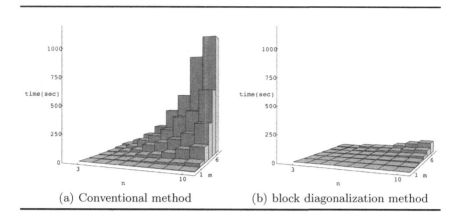

(a) Conventional method   (b) block diagonalization method

FIGURE 6.33
Required computation time for various n, m

6.4.5   Evaluation of Computational Efficiency of Real Eigenvalue Analysis

Using a workstation NWS–1750 made by Sony, a program was created by FORTRAN, and the eigenvalue analysis of the matrix A was performed under the same conditions as much as possible. Numerical analysis by this method consists of reading coordinate transformation matrix H, coordinate transformation of matrices M, K, and C by formula (6.17), inverse matrix (computation of $M^\mu)^{-1}$, construction of unsymmetric matrix A^μ, balancing to obtain a stable numerical solution, conversion of matrix A to Hessenberg matrix, and QR iteration. It consists of the analysis of the eigenvalues and eigenvectors of the asymmetric matrix A. Note that the computer environment at this time is the TSS server environment, so it is not a single task.

The computation times required for the conventional eigenvalue analysis method and the proposed method when the parameters n and m are varied are obtained, **Fig. 6.33**(a) and (b), respectively. For example, when $n = 10$, $m = 6$, the analysis time required by this method is reduced to about $1/10$ compared to the conventional method, and the block diagonalization method is clearly computationally efficient. As shown in the formula (6.61), the computational efficiency is expected to improve further as n increases.

Plotting the size N of matrices K, C, and M as well as the analysis time (sec) on a log-log graph yields **Fig. 6.34**. The straight line in the figure represents the regression line, the ($\bullet$) mark is the data of the conventional method, and ($\square$) and ($\blacksquare$) are the data of the present method. In particular, the ($\blacksquare$) mark is data that satisfies $n = m$ ($c = 1$ in the formula (6.62)). The regression line of this method was found for the data satisfying $n = m$. The slope of the data for the conventional method is approximately 3, and 2 for the

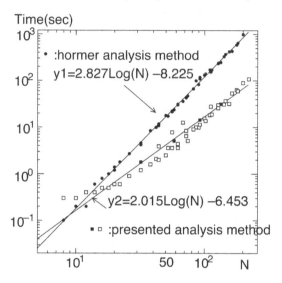

FIGURE 6.34
Comparison of QR iteration times for matrix size N

present method. This means that the computational cost of the conventional method is approximately proportional to N^3, and the cost of this method is approximately proportional to N^2. This result confirms the validity of the computational cost derived based on the assumption (6.62) and the evaluation formula (6.63) for the ratio of computation time between the conventional method and the proposed method. The other data ($\square$) also roughly follow the regression line, and the above discussion can be considered to hold as it is.

7

Numerical Efficiency Evaluation of Parallel Computing Method

A method of block diagonalization (BDM) such as the stiffness matrix, the damping matrix, and the mass matrix of symmetric structures has been proposed, and its numerical advantages have been shown [17]- [30]. This method decomposes the stiffness matrix of a symmetric system into multiple independent equations by using a coordinate transformation based on its geometric characteristics.

7.1 Numerical Efficiency Evaluation of Parallel Cholesky Decomposition for Stiffness Matrices of Symmetric Structures via Block Diagonalization Method

This section aims to apply the BDM to the modified Cholesky method, commonly used for matrix triangular decomposition, and attempts to convert the upper and lower triangular matrices themselves into upper and lower triangular forms in diagonal block units. That is, we propose a parallel computation method for *LU* decomposition that fully orthogonally decomposes upper and lower triangular matrices into multiple independent upper and lower triangular small block matrices via a coordinate transformation based on geometric characteristics. We will then examine numerical analysis evaluation methods for this computation method. By this computation method, the size of the matrices used for numerical computations can be significantly reduced, achieving both an improvement in computational efficiency and a realization of parallel analysis for numerical analysis.

Furthermore, we calculated an efficiency evaluation formula for operations on small matrices orthogonalized by BDM, and a calculation error evaluation formula based on condition numbers. In addition, we objectively compared the operational efficiency and the required array capacity associated with the parallelization of *LU* decomposition with the direct method, using regression equations, among other things. As a numerical analysis example, we apply this method to the dome structure of a symmetric structure system (a system

DOI: 10.1201/9781032670386-7

covariant with the Dihedral group), perform a numerical analysis evaluation, and verify the usefulness and validity of this method.

7.1.1   Parallel Cholesky Decomposition

Consider a method to parallelize the LU decomposition of a regular matrix K into a product of lower triangular matrix L and upper triangular matrix U. Especially, when these two triangular matrices are in a transposed matrix relationship, it is called Cholesky decomposition. In this paper, we will use the modified Cholesky decomposition method, which further decomposes into $K = U^{\mathrm{T}} D U$.[1] When the stiffness matrix K^{μ} in formula (5.17) is decomposed by the modified Cholesky method, it becomes

$$K^{\mu} = (U^{\mu})^{\mathrm{T}} D^{\mu} U^{\mu}, \quad {}^{\forall}\mu \in R(G) \tag{7.1}$$

Here, U^{μ}, D^{μ} are the upper triangular matrix and diagonal matrix for each irreducible representation.

On the other hand, the block diagonal matrix $\widetilde{K}$ can be represented as

$$
\begin{aligned}
\widetilde{K} &= \operatorname{diag}[\cdots, (H^{\mu})^{\mathrm{T}} K H^{\mu}, \cdots] \\
&= \operatorname{diag}[\cdots, K^{\mu}, \cdots] \\
&= \operatorname{diag}[\cdots, (U^{\mu})^{\mathrm{T}} D^{\mu} U^{\mu}, \cdots] \\
&= \widetilde{U}^{\mathrm{T}} \widetilde{D} \widetilde{U}
\end{aligned}
\tag{7.2}
$$

from formulas (3.10) and (7.1). Here,

$$
\begin{aligned}
\widetilde{U} &= \operatorname{diag}[\cdots, U^{\mu}, \cdots] \\
\widetilde{D} &= \operatorname{diag}[\cdots, D^{\mu}, \cdots]
\end{aligned}
$$

decomposition is possible for the irreducible representation μ.

For instance, if we write the specific form of this $\widetilde{U}$, we obtain a block triangularized matrix form as follows:

Thus, since each block is completely independent, operations can be performed for each block, which is also suitable for high-speed calculation by parallel computers. Or even on a single machine, the matrix size of each block is reduced, enabling saving of main memory and higher speed.

---

[1] There are methods such as LU decomposition and block modified Cholesky decomposition, often used when the performance of the main memory causes the expanded matrix to be divided into rows and columns, which are called blocks. However, this method is based on a completely different idea from these.

7.2 Evaluation of Computational Performance

In this chapter, we compare the analysis efficiency of obtaining the numerical solution of the stiffness equations, which are simultaneous equations, between the conventional direct method and BDM.

7.2.1 Evaluation of Computational Efficiency

When considering the computational efficiency of simultaneous equations, the efficiency of matrix operations is governed by the number of iterations and the degrees of freedom, but in this paper, we focus on the degrees of freedom (matrix size) N, which is the most dominant in terms of required array capacity and analysis time.

Assume that the main term of matrix operations (or array capacity) ρ by the direct method is expressed as

$$\rho \equiv aN^\alpha \qquad (7.3)$$

where a and α are constants. In this case, the main term of the operation amount (or array amount) by BDM is

$$\widetilde{\rho} = \sum_{\mu \in R(G)} a(N^\mu)^\alpha \qquad (7.4)$$

where N^μ represents the matrix size of the irreducible representation μ. Also, in a computer capable of parallel computation with the number of irreducible representations, the maximum matrix size in the matrices of the irreducible representations becomes dominant in the computation time.

$$\widetilde{\rho}' = \mathrm{Max}(\cdots, a(N^\mu)^\alpha, \cdots). \qquad (7.5)$$

In other words, the operation amount for this matrix becomes dominant. The operation amount (and array amount) associated with the coordinate transformation of Eq.(3.11) can be made negligible compared to this cost by using local transformations [31], so we ignored it in Eq.(7.5).

Assuming that the size N^μ of the block K^μ is all the same, [2] we can approximate

$$N^\mu \simeq \frac{N}{q(G)} \qquad (7.6)$$

where $q(G)$ represents the total number of blocks of the group G. In this

---

[2]The size of the block matrix varies depending on the group, but in the case of a Dihedral group, the second-order irreducible representation is generally constant.

case, by substituting Eq.(7.6) into Eq.(7.5), the operation amount (or array amount) by BDM can be approximated as

$$\tilde{\rho} \simeq a \left(\frac{N^{\alpha}}{[q(G)]^{\alpha-1}} \right) \tag{7.7}$$

and the ratio of the operation amount (or array capacity) of this method to the conventional method can be approximated as

$$\frac{\tilde{\rho}}{\rho} \propto \frac{1}{[q(G)]^{\alpha-1}} \tag{7.8}$$

from equations (7.3) and (7.7). Furthermore, in a parallel computer, the operation amount and calculation efficiency can be improved to

$$\tilde{\rho}' \simeq a \left(\frac{N}{q(G)} \right)^{\alpha}, \quad \frac{\tilde{\rho}'}{\rho} \propto \frac{1}{[q(G)]^{\alpha}} \tag{7.9}$$

and the superiority is clear. These equations indicate that the computational efficiency of this method improves as the number of blocks $q(G)$ increases for $\alpha \geq 2$.

7.3   Pseudo-Diagonal Block Transformation

In recent years, a method known as Block Diagonalization Method (BDM) has been utilized to diagonalize the stiffness matrices of symmetric structural systems [17, 46]. This method performs a coordinate transformation of the stiffness matrix of a symmetric system into a coordinate system that spans its geometrical symmetry, effectively parallelizing the stiffness equations. For example, structural systems with symmetry close to circles or spheres have high-order symmetry, which is used as valuable geometric information in engineering applications. However, so far, its usage has been limited to the global symmetry of the entire structure, and it has not been extended to the use of locally different symmetries.

In this research, we propose a new computational algorithm and a further application of symmetry by applying our method to structural analysis of structures with multiple different symmetries. By performing coordinate transformations on the stiffness matrices of locally symmetric structures that correspond to the local symmetry of each orbit, a pseudo-diagonal block matrix with a phase structure is generated. Furthermore, this block structure is formed from the relationship of hierarchical symmetry connections (known as a in applied mathematics), and diagonally block structures with a genetic hierarchical structure appear. As a result, we developed an algorithm that parallelizes the iterative calculation of the stiffness equations of locally symmetric structures. After transforming into a pseudo-diagonal block, apply the

Successive Over-Relaxation method (Gauss elimination method is also possible) in parallel for each diagonal block, realizing both the parallelization and acceleration of numerical calculations and the reduction of the required array capacity. As a numerical example, the method is applied to the net structure of symmetric structural systems, verifying the usefulness and validity of the method.

We discuss the transformation theory for the pseudo-diagonal blockization of the stiffness equations of structures with local symmetries. Let's assume the stiffness equation of a locally symmetric structure with multiple G_i invariant substructures as follows:

$$F(u, f) = Ku - f = 0 \qquad (7.10)$$

Here, u and f represent the displacement vector and the external force vector, respectively, while K denotes the stiffness matrix.

$$K = \begin{pmatrix} K_{G_1} & K_{G_1 G_2} & & & O \\ K_{G_2 G_1}^{\mathrm{T}} & \ddots & & \ddots & \\ & & \ddots & K_{G_{M-1}} & K_{G_{M-1} G_M} \\ O & & & K_{G_M G_{M-1}}^{\mathrm{T}} & K_{G_M} \end{pmatrix} \qquad (7.11)$$

The stiffness equation takes into account local symmetries. In this paper, we focus on structures with increasing local symmetry. [3] The symmetry group G_i of the substructure, composed of elements g representing geometric transformations, has a chain adaptation relationship of subgroups:

$$G_1 \subset G_2 \subset \cdots \subset G_i \subset G_{i+1} \subset \cdots \subset G_M \qquad (7.12)$$

Here, G_M is the symmetry group of the entire structure. This chain adaptation relation is given by the increasing symmetry of the substructures. In the above relationship, the left side of $\subset$ represents the parent group G_i, and the right side represents the proper subgroup of G_i. The overall symmetry of the structure is defined by the group G. This is well-known from group representation theory. Here, we assume that the group element g of the symmetry group of the substructure satisfies the group action transformation

$$T_i(g)u = g(u), \quad {}^{\forall}g \in G \qquad (7.13)$$

where $T_i(g)$ is the $N \times N$ representation matrix of the group element g. In this context, we consider the force vector f to exist in the same space as $u \in \mathbf{R}^N$.

The local symmetry of this system is given by the equation F's covariance condition, which is the invariance of the equation F with respect to the mapping transformation $T_i(g)$ under the action of the element g of the group G_i,

$$T_i(g)F(u, f) = F(T_i(g)u, T_i(g)f), \quad {}^{\forall}g \in G \qquad (7.14)$$

---

[3] Although we are considering structures with increasing symmetry here, this theory can easily be applied to structures with decreasing symmetry.

This is a generalized form of the geometric symmetry condition. From the fact that this equation holds for all $\boldsymbol{u}$, we can obtain the condition

$$T_i(g)K_{G_i} = K_{G_i}T_i(g), \quad {}^{\forall}g \in G \tag{7.15}$$

This indicates that the substructure K_{G_i} can be block-diagonalized by a certain coordinate transformation matrix. The coordinate transformation matrix is a unitary matrix composed of this irreducible representation $\mu_i \in R(G_i)$ [27] ~ [31].

Furthermore, let the set of all irreducible representations of the symmetry group of the substructures $\{G_i \in G | G_1, \cdots, G_M\}$ be $R(G_i)$, and let the irreducible representation μ_{i+1} restricted from group G_{i+1} to G_i be denoted as $\mu_{i+1} \downarrow G_i$ [63]. Such a restricted representation does not necessarily have to be irreducible, and from the chain-adaptable basis, it is represented as

$$\mu_{i+1} \downarrow G_i = \sum_{\mu_i \in R(G_i)} m_{\mu_i}^{\mu_{i+1}} \cdot \mu_i, \quad {}^{\forall}\mu_{i+1} \in R(G_{i+1}) \tag{7.16}$$

where $m_{\mu_i}^{\mu_{i+1}}$ is a natural number. This systematically classifies the chain-adaptable relationships between the representations derived from different symmetry groups, representing the mechanism of hierarchical substructures and enabling parallel computation by pseudo-diagonal blocks.

7.4   Parallel Iterative Algorithm for Pseudo-Diagonal Blocks

We apply the principle of parallelizing iterative calculations for pseudo-diagonal blocks to the SOR method, which has high convergence properties for linear iteration. This method is attempted for linear iteration and preconditioned iterative methods. After block-diagonalizing the stiffness equations of a locally symmetric structural system, we define the pseudo-diagonal block matrix that emerges as a strictly superior diagonal block matrix in the following:

$$\phi_i \equiv \sum_{j=1}^{n^{\mu_i}} |\widetilde{K}^{\mu_i}{}_{ij}| - \sum_{\substack{\mu_j \in R(G_j) \\ \mu_j \neq \mu_i}} \sum_{j=1}^{n^{\mu_i}} |\widehat{K}^{\mu_i}_{\mu_j}{}_{ij}| > 0,$$

$$i = 1, 2, \cdots n^{\mu_i}, \quad \mu_i \in R(G_i)$$

$$\phi_j \equiv \sum_{i=1}^{n^{\mu_j}} |\widetilde{K}^{\mu_j}{}_{ij}| - \sum_{\substack{\mu_i \in R(G_i) \\ \mu_i \neq \mu_j}} \sum_{i=1}^{n^{\mu_j}} |\widehat{K}^{\mu_j}_{\mu_i}{}_{ij}| > 0,$$

$$j = 1, 2, \cdots n^{\mu_j}, \quad \mu_j \in R(G_j) \tag{7.17}$$

Furthermore, each diagonal block is defined as a dominant matrix with respect to the diagonal elements.

$$|\widehat{K}^{\mu_i}{}_{ii}| > \sum_{\substack{j=1 \\ j \neq i}}^{n^{\mu_i}} |\widehat{K}^{\mu_i}{}_{ij}|, \quad i = 1, 2, \cdots, n^{\mu_i}, \quad \mu_i \in R(G_i)$$

$$|\widehat{K}^{\mu}_{jj}| > \sum_{\substack{i=1 \\ i \neq j}}^{n^{\mu}} |\widehat{K}^{\mu_j}{}_{ij}|, \quad j = 1, 2, \cdots, n^{\mu_j}, \quad \mu_j \in R(G_j)$$

where, n^{μ_i}, n^{μ_j} represent the matrix size of a certain block matrix. This matrix is characterized by dominant diagonal (block) elements, and it is said to exhibit good convergence stability when this condition is satisfied.

The pseudo-diagonal block matrix $\widetilde{K}^{\mu}$ is represented as the sum of the diagonal block $\widehat{K}^{\mu}$ and the off-diagonal block Q:

$$\widetilde{K}^{\mu} = \widehat{K}^{\mu} + Q \tag{7.18}$$

where,

$$\widehat{K}^{\mu} = \text{diag}\left[\widehat{K}^{\mu_1}, \widehat{K}^{\mu_2}, \cdots, \widehat{K}^{\mu_m}\right] \tag{7.19}$$

The equation for the pseudo-diagonal block matrix is represented as

$$\left(\widehat{K}^{\mu} + Q\right) \boldsymbol{w}^{\mu} - \boldsymbol{b}^{\mu} = \mathbf{0} \tag{7.20}$$

where $\boldsymbol{w}^{\mu} = (H^{\mu})^{\mathrm{T}}\boldsymbol{u}$, $\boldsymbol{b}^{\mu} = (H^{\mu})^{\mathrm{T}}\boldsymbol{f}$. We obtain the general linear iterative equation

$$\boldsymbol{w}^{\mu}_{k+1} = (\widehat{K}^{\mu})^{-1}\boldsymbol{r}^{\mu} \tag{7.21}$$

Here, $\boldsymbol{r}^{\mu}$ is the residual vector $(\boldsymbol{r}^{\mu} = \boldsymbol{b}^{\mu} - Q\boldsymbol{w}^{\mu}_k)$. That is,

$$\boldsymbol{w}^{\mu_i}_{k+1} = (\widehat{K}^{\mu_i})^{-1}\left((H^{\mu_i})^{\mathrm{T}}\boldsymbol{f} - \sum_{\substack{\mu_j \in R(G_j) \\ \mu_j \neq \mu_i}} \widehat{K}^{\mu_i}_{\mu_j}\boldsymbol{w}^{\mu_j}_k\right),$$

$$^{\forall}\mu_i \in R(G_i), \quad k = 1, 2, \cdots \tag{7.22}$$

is obtained. When the irreducible dominant diagonal block $\widehat{K}^{\mu}$ is sufficiently dominant over the solution to ignore the off-diagonal blocks, rough approximate solutions or initial values for iterative calculations can be obtained. By setting $\boldsymbol{w}^{\mu_j}_0 = \mathbf{0}$ in Eq.(7.22),

$$\boldsymbol{w}^{\mu}_1 = (\widehat{K}^{\mu})^{-1}\boldsymbol{b}^{\mu} = \begin{pmatrix} (\widehat{K}^{\mu_1})^{-1}\boldsymbol{b}^{\mu_1} \\ \vdots \\ (\widehat{K}^{\mu_m})^{-1}\boldsymbol{b}^{\mu_m} \end{pmatrix} \tag{7.23}$$

A simplified first approximation solution can be obtained.

By the way, if $\widehat{K}^\mu$ is the sum of the lower triangular matrix, diagonal matrix, and upper triangular matrix for each irreducible representation,

$$\widehat{K}^\mu = \widehat{L}^\mu + \widehat{D}^\mu + \widehat{R}^\mu, \quad {}^\forall \mu \in R(G) \tag{7.24}$$

then Eq.(7.22) can be solved as

$$
\begin{aligned}
\boldsymbol{w}^\mu_{k+1} &= \omega(\widetilde{D}^\mu)^{-1}\left(\boldsymbol{r}^\mu - \widetilde{L}^\mu \boldsymbol{w}^\mu_{k+1} - \widetilde{R}^\mu \boldsymbol{w}^\mu_k\right) \\
&\quad + (1-\omega)\boldsymbol{w}^\mu_k, \quad {}^\forall \mu \in R(G)
\end{aligned}
\tag{7.25}
$$

Additionally, since $\widehat{K}^{\mu_i}_{\mu_j}$ of the residual vector $\boldsymbol{r}^\mu$ can be decomposed into the lower triangular matrix $\widehat{L}$ and the upper triangular matrix $\widehat{R}$, it can be expressed as

$$
\begin{aligned}
\boldsymbol{r}^\mu &= \boldsymbol{b}^\mu - \sum_{\substack{\mu_j \in R(G_j) \\ \mu_j \neq \mu_i}} \widehat{K}^{\mu_i}_{\mu_j} \boldsymbol{w}^\mu_k \\
&= \boldsymbol{b}^\mu - \widehat{L} \boldsymbol{w}^\mu_{k+1} - \widehat{R} \boldsymbol{w}^\mu_k \\
&= \boldsymbol{b}^\mu - \sum_{\substack{\mu_j \in R(G_j) \\ \mu_j < \mu_i}} \widehat{K}^{\mu_i}_{\mu_j} \boldsymbol{w}^{\mu_j}_{k+1} - \sum_{\substack{\mu_j \in R(G_j) \\ \mu_j > \mu_i}} \widehat{K}^{\mu_i}_{\mu_j} \boldsymbol{w}^{\mu_j}_k, \quad {}^\forall \mu \in R(G) \tag{7.26}
\end{aligned}
$$

Furthermore, if $\widehat{K}^\mu$ is a symmetric matrix, only one of the triangular matrices is sufficient. This series of calculations involves calculating approximate solutions in parallel for each irreducible representation μ_i, and then iteratively converging for $k = 1, 2, \cdots$. This method fundamentally differs from the conventional block iterative method (Successive Block Over-Relaxation method), which is a block partitioning method due to the constraints of traditional computers. It utilizes the orthogonality of the diagonal block matrix for each irreducible representation. As all terms apart from the correction term are completely independent, parallel computation is possible.

7.5 Numerical Calculation of Pseudo-Diagonal Blocks

Let's consider the pseudo-diagonal block transformation of the stiffness equation of the locally symmetric network structure shown in **Fig. 7.1**. The orbit of this structural model is set to $m = 3$, and the degree of freedom is 1 degree of freedom/node in the Z direction displacement. **Fig. 7.2** shows an image of the entire stiffness matrix consisting of one node (one degree of freedom in the Z direction) of the structural model ($n = 4, m = 3$). For instance, there exist six irreducible representations of the D$_4$ structural system [46]:

$$\mu = (1,1)_{D_4}, (1,2)_{D_4}, (1,3)_{D_4}, (1,4)_{D_4}, (2,1)^+_{D_4}, (2,1)^-_{D_4}$$

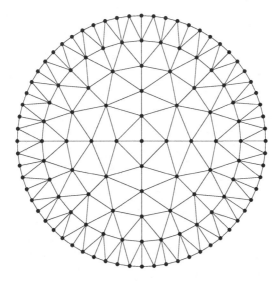

FIGURE 7.1
D_n invariant system m-layer network structure

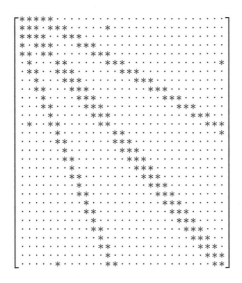

FIGURE 7.2
Global Stiffness Matrix K

```
⎡ A A · · · · · · · · · · · · · · · · · · · · · · · ⎤
⎢ A A · · A A · · · · · · · · · · · · · · · · · · · ⎥
⎢ · · B · · · · · B · · · · · · · · · · · · · · · · ⎥
⎢ · · C · · C · C · · · · · · · · · · · · · · · · · ⎥
⎢ · · · D · · · · · D · D · · · · · · · · · · · · · ⎥
⎢ · A · · · A A · · · · · · A A · · · · · · · · · · ⎥
⎢ · A · · · A A · · · · · · · · · A · · · · · · · · ⎥
⎢ · · · C · · C · C · · · · · C · · · · C · · · · · ⎥
⎢ · · B · · · · · B · · · · · B · · B · · · · · · · ⎥
⎢ · · · C · · C · C · · D · · · · C · C · · · · · · ⎥
⎢ · · · · D · · · · · D · D · · · · · · · D · · · · D ⎥
⎢ · · · · · · · · E · · · · · · · · · · E · · · E · ⎥
⎢ · · · · D · · · · · D · D · · · · · · · D · D · · ⎥
⎢ · · · · · A · · · · · · A A · · · · · · · · · · · ⎥
⎢ · · · · · A · · · · · · A A · · · · · · · · · · · ⎥
⎢ · · · · · · C · · · · · · · C · · · · C · · · · · ⎥
⎢ · · · · · · · B · · · · · · B · · B · · · · · · · ⎥
⎢ · · · · · · · C · · · · · · · C · C · · · · · · · ⎥
⎢ · · · · · A · · · · · · · · · A · · · · · · · · · ⎥
⎢ · · · · · · · C · · · · · · · C · C · · · · · · · ⎥
⎢ · · · · · · · B · · · · · · B · · · B · · · · · · ⎥
⎢ · · · · · · · C · · · · · · · C · · · · C · · · · ⎥
⎢ · · · · · · · · D · · · · · · · · · D · · · · · D ⎥
⎢ · · · · · · · · · E · · · · · · · · · · E · · · E · ⎥
⎢ · · · · · · · · · D · · · · · · · · · · D · D · · ⎥
⎢ · · · · · · · · · · · · · · · · · · · · · · F · · · ⎥
⎢ · · · · · · · · · D · · · · · · · · · · D · D · · ⎥
⎢ · · · · · · · · · E · · · · · · · · · · E · · · E · ⎥
⎣ · · · · · · · · · D · · · · · · · · · · D · · · · D ⎦
```

FIGURE 7.3
Direct coordinate transformation

Using the coordinate transformation matrices corresponding to each symmetry, such as

$$H = \begin{bmatrix} H_{G_1} & & O \\ & H_{G_2} & \\ O & & H_{G_3} \end{bmatrix}, \quad \begin{cases} G_1 = D_4 \\ G_2 = D_8 \\ G_3 = D_{16} \end{cases}$$

When the coordinate transformation consisting of the irreducible representation matrix is performed directly from the inner orbit,

$$\widetilde{K} = H^{\mathrm{T}} K H$$

then, it is transformed into a matrix image as shown in **Fig. 7.3**. After sorting, the matrix can be completely transformed into a block diagonal form as shown in **Fig. 7.4**. The blocks A, B, C, D, E, and F belong to the irreducible representations $(1,1)_{D_4}$, $(1,3)_{D_4}$, $(2,1)^{+}_{D_4}$, $(2,1)^{-}_{D_4}$, $(1,2)_{D_4}$ and $(1,4)_{D_4}$, respectively. If you look closely at this diagonal, you will find that the higher-order symmetry where the orbital is located on the outer side is further derived into orthogonal small-block matrices. This suggests that genetic parallel computation is possible with further strand adaptation.

For instance, the block C (i.e., $\widehat{K}^{(2,1)^{+}_{D_4}}$) further forms quasi-diagonal block cells. When represented within this block using irreducible representations, the

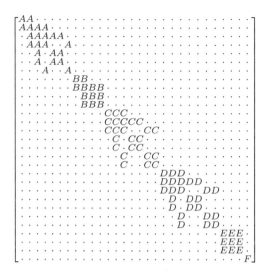

FIGURE 7.4
Transformation to block diagonal form

$$\mu_1 \left\{ \begin{array}{c} \mu_2 \\ \mu_3 \end{array} \right.$$

FIGURE 7.5
Hierarchical structure of irreducible expressions μ_1, μ_2, μ_3

following equation holds:

$$\begin{bmatrix} \widehat{K}^{\mu_1} & \widehat{K}^{\mu_1}_{\mu_2} & \widehat{K}^{\mu_1}_{\mu_3} \\ \widehat{K}^{\mu_2}_{\mu_1} & \widehat{K}^{\mu_2} & O \\ \widehat{K}^{\mu_3}_{\mu_1} & O & \widehat{K}^{\mu_3} \end{bmatrix} \left\{ \begin{array}{c} w^{\mu_1} \\ w^{\mu_2} \\ w^{\mu_3} \end{array} \right\} = b^{\mu} \tag{7.27}$$

The chain compatibility relation between these irreducible representations is shown in **Fig. 7.5**, and after the computation of the small block μ_1, the computations of μ_2 and μ_3 are parallel computation independent of each other is possible. Based on this computation principle, iterative calculations are locally parallelized.

7.5.1   Numerical Example by Quasi-block-diagonal Method

The normalized irreducible dominant diagonal block $\widehat{K}^{\mu}$ in Eq.(7.27) is given by

$$\widehat{K}^{\mu} = \mathrm{diag}\left[\widehat{K}^{\mu_1}, \widehat{K}^{\mu_2}, \widehat{K}^{\mu_3}\right] \tag{7.28}$$

Here, the components when $EI/\ell^3 = 1$ are normalized are as follows.

$$\widehat{K}^{\mu_1} = \begin{bmatrix} 3.3 & -1.0 & -0.4 \\ -1.0 & 2.7 & 0.2 \\ -0.4 & 0.2 & 3.5 \end{bmatrix},$$

$$\widehat{K}^{\mu_2} = \begin{bmatrix} 2.0 & 0.1 \\ 0.1 & 3.3 \end{bmatrix}, \quad \widehat{K}^{\mu_3} = \begin{bmatrix} 1.7 & 0.1 \\ 0.1 & 2.6 \end{bmatrix}$$

are obtained. The components of the off-diagonal block matrix are as follows.

$$\widehat{K}^{\mu_1}_{\mu_2} = \begin{bmatrix} 0 & 0 \\ -1.2 & -0.2 \\ 0 & 0 \end{bmatrix}, \quad \widehat{K}^{\mu_1}_{\mu_3} = \begin{bmatrix} 0 & 0 \\ 0 & 0 \\ -0.9 & -0.5 \end{bmatrix}$$

Let's consider the numerical calculation of this pseudodiagonal block after block diagonalization. We verify whether the convergence stability of this block is ensured by examining ϕ_i $(i = 1, 2, \cdots)$ as per Eq.(7.17). The result is

$$\phi_i > 0$$

indicating that it is a strict irreducible dominant diagonal block. Moreover, when the condition number is calculated using the Z condition number, we get

$$Z(\widehat{K}^{\mu}) = ||(\widehat{K}^{\mu})^{-1}||_{\infty} \cdot ||Q||_{\infty} = 0.856$$

With the relative error ratio of the matrix norm being $||\delta w||/||w|| \leq 5.94$, we cannot ignore the off-diagonal blocks of this pseudodiagonal block matrix.

By the way, let's set the load vector after the transformation on the right-hand side as

$$b^{(2,1)^+_4} = 10\,(1, 1, 1, 1, 1, 1, 1)^{\mathrm{T}}$$

The results of the number of convergence iterations and the error by the proposed method are shown in **Fig. 7.6**. A numerical solution was obtained at the 5th iteration.

The real parallel calculation is possible after calculating the solution w^{μ_1} for block $\widehat{K}^{\mu_1}$, at the calculation stage of blocks $\widehat{K}^{\mu_2}$ and $\widehat{K}^{\mu_3}$ the solutions w^{μ_2} and w^{μ_3} have a linearly independent relationship. Therefore, locally independent iterative calculations are possible. That is, the irreducible representations of this structural system have a chain adaptation relationship as shown in **Fig. 7.7**, and after calculating block $(2, 1)^+_{D_4}$, representations below $(2, 1)^+_{D_8}$ and below $(2, 3)^+_{D_8}$ become equal independent spaces.

TABLE 7.1

Numerical calculation results using parallel SOR method

No.	$\|r\|$	w^{μ_1}			w^{μ_2}		w^{μ_3}	
1	14.6477	3.030	3.704	2.857	5.000	3.030	5.882	3.846
2	0.4262	4.793	8.516	5.405	10.950	3.251	9.045	4.676
3	0.0335	6.561	11.305	6.065	11.755	3.381	8.773	4.675
4	0.0006	7.317	11.466	5.941	11.702	3.369	8.749	4.648
5	0.0000	7.207	11.366	5.946	11.641	3.366	8.758	4.654
6	0.0000	7.193	11.348	5.948	11.640	3.365	8.757	4.653
7	0.0000	7.189	11.349	5.947	11.641	3.365	8.757	4.653
Gauss	——	7.190	11.350	5.947	11.641	3.365	8.757	4.653

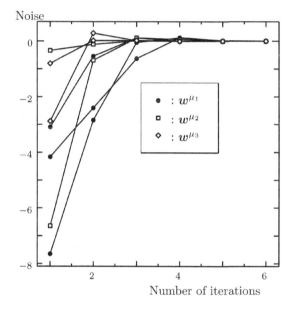

FIGURE 7.6

Convergence iterations and error using the proposed method

$$(2,1)^+_{D_4}\begin{cases}(2,1)^+_{D_8}\begin{cases}(2,1)^+_{D_{16}}\\(2,7)^+_{D_{16}}\end{cases}\\(2,3)^+_{D_8}\begin{cases}(2,5)^+_{D_{16}}\\(2,3)^+_{D_{16}}\end{cases}\end{cases}$$

FIGURE 7.7

Hierarchical structure of the irreducible representation $(2,1)^+_{D_m}$ for D_4, D_8, and D_{16} ($m=3$)

TABLE 7.2
Numerical calculation results of parallel SOR method (in μsec)

Method	w^{μ_1}	w^{μ_2}	w^{μ_3}	Total
SOR Method	117	76	76	269
Parallel SOR method	117	76		193
Gauss method	86			86

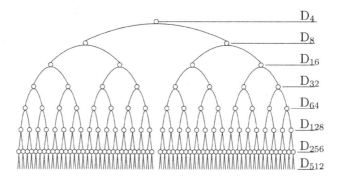

FIGURE 7.8
Chain adaptation structure diagram of $\mu = (2,1)^+_{D_4}$ in D_{128} structure system

    This method enables local parallel calculation from the hierarchical structure of symmetry. Also, the calculation time in these matrices was measured under the same conditions as much as possible. The results are shown in **Table 7.2**. In this model, the direct method, Gauss elimination method obtained 86μsec, the SOR iterative method obtained 269μsec, and the proposed method obtained 193 μsec. In this case, the parallel took 2.2 times the calculation time of the direct method. For such small-scale structures, the calculation by the direct method (Gauss elimination method) yielded superior results. Next, let's consider a large-scale local symmetry structure system ($n = 128, m = 6$) based on these computational principles. The overall symmetry of this system is under the action of the D_4 group, so there are six irreducible representations of this group: $(1,1)_{D_4}, (1,2)_{D_4}, (1,3)_{D_4}, (1,4)_{D_4}, (2,1)^+_{D_4}, (2,1)^-_{D_4}$. Among these, the chain-adapted base and hierarchical structure of symmetry for $(2,1)^+_{D_4}$ are shown in **Fig. 7.8**. This hierarchical structure possesses a typical doubling periodic phase structure. By representing the symmetry of each of these orbits with the chain-adapted base of the irreducible representation, and expressing it in matrix form from lower-order symmetries, we obtain a pseudo-diagonal block form as shown in **Fig. 7.9**. The form of this matrix is fundamentally different from the matrix space like **Fig. 7.2**. If this matrix were a symmetric matrix, as shown in **Fig. 7.10**, we would only need to secure the non-zero

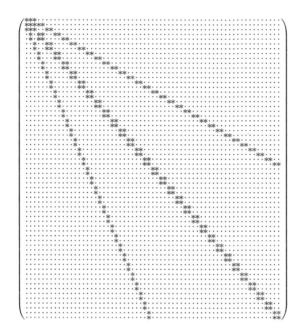

FIGURE 7.9
Pseudo-diagonal block matrix for the $\mu = (2,1)_{D_4}^+$ of the D_{128} structural system

block components. In this way, the reduced matrix size can be compressed to about 8% of the entire pseudo-diagonal block. Therefore, we can expect even more compression effects for the entire stiffness matrix.

Finally, we illustrate in **Fig. 7.11** the computation time for different methods when the parameter m in **Fig. 7.1** is varied. The symbols $\square$, $\circ$, and $\bullet$ represent the computation times when using Gauss's elimination method, the SOR method of linear iteration, and our method, respectively. Although direct methods were superior for small-scale structure systems, for the large-scale D_{128} system, this method finished in about 1/10th of the computation time of the direct method. Since these calculations are all within one irreducible representation of the quasi-diagonal block form, comparing with the

FIGURE 7.10
Example of array matrix storage for the D_{128} structure system

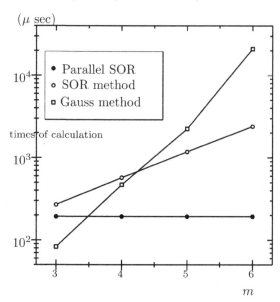

FIGURE 7.11
Comparison of computation times for various methods against parameter m

calculation of the full band matrix of the stiffness matrix before direct transformation, a stark difference will emerge.

In this study, we were able to demonstrate a computational principle that
enables the parallel acceleration of linear iteration methods for the stiffness
equations of locally symmetric structure systems, by using pseudo-diagonal
blocks.

8

Products of Group Representation

This chapter attempts to describe and utilize higher-order symmetry in element stiffness matrices on two- and three-dimensional spaces, using the standard decomposition theorem for the $D_\infty \times D_\infty$ invariant system by Ikeda and Murota [69]. Specifically, this chapter focuses on the transformation rules of low-dimensional symmetries, expresses the high-dimensional symmetries assembled from them, and proposes a way of thinking and compatibility of symmetries for each structural member. Specifically, in the case of a three-dimensional cuboidal solid element, we aim to orthogonalize the block matrices of the stiffness matrix using combinations (group products) of basic symmetry transformations, and examine the validity and effectiveness of this decomposition method. In the structural analysis examples, we focus on analytical examples that are prone to problems in the structural analysis of symmetrical structures, as problems to understand the mechanism of high-order symmetry. In this way, the use of group products is important in understanding the mechanism of the symmetry group that a structure has.

8.1 Symmetry of a Cuboid

In this section, we will describe the symmetry operations of the geometrical symmetry of a three-dimensional cuboid. Finite element analysis with solid elements has versatility in terms of its application to design models of real structures, but it carries problems leading to large-scale structural calculations [64, 67, 68]. Recently, an 8-node solid element that can be applied to various element members has been developed, and good analysis results have been obtained. However, in general, even a solid rectangular element without rotational degrees of freedom forms a 24×24 size full element stiffness matrix, leading to problems such as increased computational cost with the increase in discretized elements and variations in computational accuracy. As one way to solve these problems, for example, methods such as foldback symmetry and domain decomposition have been used, but these methods utilized only a limited symmetry.

As a way of describing symmetry, it is standard to use groups that express symmetry [12, 14]. In particular, the utilization of symmetry is established

DOI: 10.1201/9781032670386-8

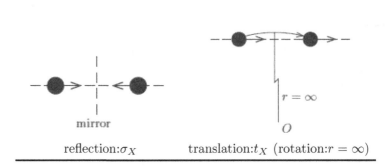

FIGURE 8.1
Source for left-right symmetry

because the governing equations of a symmetric system can be decomposed into several independent equations through coordinate transformation from the symmetry group of the system [15, 61]. However, in the field of structural engineering, the application of symmetry in three-dimensional finite elements has not been achieved. One reason for this is that the symmetry of three-dimensional space has higher-order symmetry, each with its own unique characteristics, and it was not easy to cover all symmetries.

Consider the cuboid region

$$\{(\xi, \eta, \zeta) \in \mathbf{R}^3 | -1 \le (\xi, \eta, \zeta) \le 1\} \tag{8.1}$$

Here, ξ, η, ζ are variables normalized by half the length of each edge. From the definition of this region, there exists a folding symmetry (mirror symmetry) with mirror planes in the X, Y, Z directions that include the origin (0,0,0). In considering the symmetry of a cuboid solid element, we will define lower-order symmetry transformations.

8.1.1 Symmetric Transformations in One-dimensional Space

We consider a transformation caused by a symmetry operation, taking only the degree of freedom in the ξ direction for a single point. The most basic symmetric transformations in one-dimensional space are mirror image transformation operations and rotation transformation operations, [1] and the transformations differ depending on the position of the symmetry operation. As shown in **Figs. 8.1 and 8.2**, we consider two transformations: one where the symmetry operation is on a line that matches the direction of the point's movement, and one where it is in a position perpendicular to it.

---

[1]The spaces for symmetry transformations and the degrees of freedom for points are generally different.

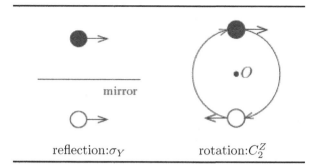

FIGURE 8.2
Source for vertical symmetry

For convenience, these will be referred to as the transformation elements for left-right symmetry and up-down symmetry, respectively. The white circles in **Fig. 8.2** represent the points after the transformation for up-down symmetry. Symmetry operations for left-right symmetry exist as

$$\sigma_X : (\xi, \eta) \mapsto (-\xi, \eta), \quad t_X : (\xi, \eta) \mapsto (\xi + \xi_0, \eta) \tag{8.2}$$

where σ_X is the mirror transformation with respect to the X direction, and t_X translates parallel in the X direction.

Similarly, the symmetry operations for up-down symmetry can be described by the transformations

$$\sigma_Y : (\xi, \eta) \mapsto (\xi, -\eta), \quad C_2^Z : (\xi, \eta) \mapsto (-\xi, -\eta) \tag{8.3}$$

where σ_Y is a mirror transformation in the Y direction and C_2^Z is an inversion transformation.

The symmetry group G_{1X} for left-right symmetry is given by

$$G_{1X} = \langle \sigma_X, t_X \rangle = \{t_X, \sigma_X t_X\}, \quad \sigma_X, t_X \in g \tag{8.4}$$

where the elements in the brackets $\langle \rangle$ are the group elements g, and G_i denotes the group corresponding to the i-dimensional space.

The symmetry operation $\sigma_X t_X$ is executed from right to left. The operation of $\sigma_X t_X$ is given by

$$\sigma_X t_X : \xi \mapsto -(\xi + \xi_0) \tag{8.5}$$

which is equivalent to a translation of $-2\xi - \xi_0$ for ξ.

Similarly, the symmetry groups G_{1Y}, G_{1Z} for up-down symmetry are given by

$$
\begin{aligned}
G_{1Y} &= \langle \sigma_Y, C_2^Z \rangle = \{C_2^Z, \sigma_Y C_2^Z\}, \quad \sigma_Y, C_2^Z \in g \\
G_{1Z} &= \langle \sigma_Z, C_2^Y \rangle = \{C_2^Y, \sigma_Z C_2^Y\}, \quad \sigma_Z, C_2^Y \in g
\end{aligned}
$$

$$\tag{8.6}$$

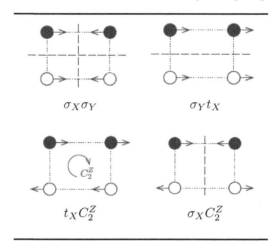

FIGURE 8.3
Symmetry of a rectangle

respectively.

8.1.2 Symmetry Transformations of Rectangles

The description of the symmetry G_2 of a rectangle with one node and one degree of freedom can be represented by the group product of the symmetry groups G_{1X} and G_{1Y} from equations (8.4) and (8.6),

$$G_2 = G_{1X} \bigotimes G_{1Y} \tag{8.7}$$

where $\bigotimes$ is the operator for group product.

This group product has a Dihedral group D_2 which shares both left-right and up-down symmetries. This group has a symmetry that fits not only rectangles but also rhombic node configurations.

The symmetry transformations of this group are composed of

$$\begin{aligned} G_2 &= \langle \sigma_X, t_X, \sigma_Y, C_2^Z \rangle \\ &= \{ \sigma_X \sigma_Y, t_X \sigma_Y, t_X C_2^Z, \sigma_X C_2^Z \} \end{aligned} \tag{8.8}$$

resulting in four types of symmetries as shown in **Fig. 8.3**. In this way, we adopt a method to grasp higher-dimensional symmetries by combining basic symmetries. [2]

---

[2]This uses the fact that the group elements composed of combinations of group elements under the same group also come under the action of that group (see Eq.(8.18)).

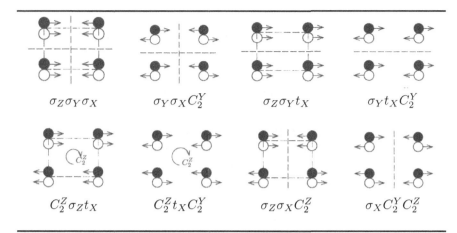

$$\sigma_Z\sigma_Y\sigma_X \qquad \sigma_Y\sigma_X C_2^Y \qquad \sigma_Z\sigma_Y t_X \qquad \sigma_Y t_X C_2^Y$$

$$C_2^Z \sigma_Z t_X \qquad C_2^Z t_X C_2^Y \qquad \sigma_Z \sigma_X C_2^Z \qquad \sigma_X C_2^Y C_2^Z$$

FIGURE 8.4
Symmetry of a cuboid (point group)

8.1.3 Symmetry Transformations of Cuboids

Similar to the previous section, the description of the symmetry G_3 of a cuboid can be represented by the group product of the $XY-$plane group G_2 and line group G_{1Z},

$$G_3 = G_2 \bigotimes G_{1Z} = G_{1X} \bigotimes G_{1Y} \bigotimes G_{1Z} \qquad (8.9)$$

This group product represents a space group consisting of Dihedral groups. [3]

This group has eight combinations of transformations, as shown in **Fig. 8.4**. That is, the symmetry transformations for a cuboid are given by

$$
\begin{aligned}
G_3 &= \left\langle \sigma_X, t_X, \sigma_Y, C_2^Z, \sigma_Z, C_2^Y \right\rangle \\
&= \{ \sigma_Y\sigma_Z\sigma_X, \sigma_Y\sigma_X C_2^Y, C_2^Z \sigma_Z t_X, C_2^Z t_X C_2^Y, \\
&\quad \sigma_Z\sigma_Y t_X, C_2^Y \sigma_Y t_X, \sigma_Z\sigma_X C_2^Z, \sigma_X C_2^Y C_2^Z \}
\end{aligned}
\qquad (8.10)
$$

It should be noted that usually, mirror symmetry (for example, the quarter-domain partitioning method) uses only the two symmetry transformations $\sigma_Z\sigma_Y\sigma_X$ and $\sigma_Y\sigma_X C_2^Y$, and does not consider the symmetry of the entire space.

Also, this group refers to elements with the spatial symmetry defined by Eq.(8.10), and the cuboid is one of its elements. Therefore, this cuboid, whether it is a truss-based framework structure, a hollow box shell structure enclosed by plate elements, or a solid element, belongs to the same group.

---

[3]This group is not limited to cuboids, but consists of points or element configurations to which the symmetry transformations apply, and is governed by point groups (for example, a regular octahedron also belongs to this group).

Using the group product in this way has the advantage of understanding the combination of symmetries of 3-dimensional elements and accurately and simplifying the use of symmetry groups.

8.1.4 Symmetry Transformations for Rotational Displacement

The symmetry for rotational displacement, as shown in **Fig. 8.5**, is different from the symmetry transformation for translational displacement. The symmetry operations for rotational displacement θ_x around the x-axis are given by

$$
\begin{aligned}
\sigma_X &: (\xi, \theta_x) \mapsto (-\xi, \theta_x), \\
C_2^Z &: (\xi, \theta_x) \mapsto (-\xi, -\theta_x), \\
\sigma_Y &: (\eta, \theta_x) \mapsto (-\eta, -\theta_x), \\
t_Y &: (\eta, \theta_x) \mapsto (\eta + \eta_0, \theta_x)
\end{aligned}
\tag{8.11}
$$

By combining these basic symmetry operations, we can represent 2D and 3D symmetry transformations in the same way as for translational displacements. The correspondence between the symmetry transformations for translational displacements and the transformations for rotational displacements is given by

$$
\begin{aligned}
\sigma_X(\xi) &\leftrightarrow \sigma_Y(\theta_x), & t_X(\xi) &\leftrightarrow t_Y(\theta_x) \\
\sigma_Y(\xi) &\leftrightarrow \sigma_X(\theta_x), & C_2^Y(\xi) &\leftrightarrow C_2^Z(\theta_x)
\end{aligned}
\tag{8.12}
$$

and they have the same symmetry transformations.

8.1.5 Degrees of Freedom in Symmetry Transformations

The symmetry transformation varies according to how the degrees of freedom of the nodes are handled, but we will consider expressing the coordinates of this space as a linear combination of each basis vector. Therefore, we consider the space spanned by these basis vectors, taking the degrees of freedom of the symmetry transformations so far as basis vectors. For example, the symmetry transformation group of a cuboid with degrees of freedom in the X, Y, Z directions is

$$
G_3(X, Y, Z) = G_3(X) \bigoplus G_3(Y) \bigoplus G_3(Z)
\tag{8.13}
$$

It can be expressed as the direct sum of the basis vectors in the X, Y, Z directions, respectively. As a result, the deformation of a cuboid can be expressed as a linear combination of symmetric deformations in the X, Y, Z directions.

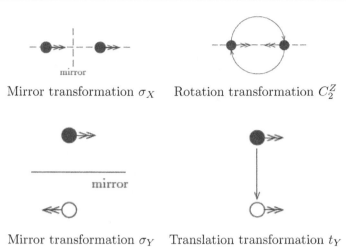

Mirror transformation σ_X Rotation transformation C_2^Z

Mirror transformation σ_Y Translation transformation t_Y

FIGURE 8.5
Transformations for rotational displacements

8.2 Coordinate Transformation Matrices of Each Symmetry Transformation

In this chapter, we will consider the mechanism of coordinate transformation matrices by expressing the symmetry transformations obtained from group products as representation matrices.

8.2.1 Representation Matrices by Group Products

The space group G_3 is a group product of the one-dimensional symmetry group G_1, and the description method of representation matrices is assumed to conform to Ikeda and Murota. The classes of non-covariant irreducible representations of G_3 ($\simeq D_2 \otimes D_2 \otimes D_2 \simeq O(3)$) are

$$R(G_3) = \{\mu \equiv +, -, 1, 2, \cdots \}. \tag{8.14}$$

Here, μ represents an irreducible representation, $+, -$ represent two types of one-dimensional irreducible representations, $1, 2, \cdots$ represent irreducible representations of order two or higher [69]. The one-dimensional first order irreducible representation matrices are expressed as

$$\begin{aligned}
T^{(+)}(t_X) &= 1, & T^{(+)}(\sigma_X) &= 1, \\
T^{(-)}(C_2^Z) &= 1, & T^{(-)}(\sigma_Y) &= -1.
\end{aligned} \tag{8.15}$$

Here, $T^{(\mu)}(\cdot)$ represents the representation matrix that expresses the action of the parent element. The representation matrices for the second-order irreducible representations are

$$T^{(n)}(t_X) = \begin{pmatrix} 1 & 0 \\ 0 & 1 \end{pmatrix}, \quad T^{(n)}(\sigma_X) = \begin{pmatrix} 1 & 0 \\ 0 & -1 \end{pmatrix},$$

$$T^{(n)}(C_2^Z) = \begin{pmatrix} -1 & 0 \\ 0 & -1 \end{pmatrix}, \quad T^{(n)}(\sigma_Y) = \begin{pmatrix} 1 & 0 \\ 0 & -1 \end{pmatrix},$$

$$n = 1, 2, \cdots \tag{8.16}$$

and the indices are

$$\chi^{(n)}(t_X) = 2, \quad \chi^{(n)}(\sigma_X) = 0,$$
$$\chi^{(n)}(C_2^Z) = -2, \quad \chi^{(n)}(\sigma_Y) = 0. \tag{8.17}$$

When there is a relationship $g_i g_j = g_k$ between any group elements under the same group, a product representation relationship also exists between the representation matrices corresponding to this,

$$T(g_i)T(g_j) = T(g_k), \quad g_i, g_j, g_k \in G \tag{8.18}$$

For example, the product representation between σ_X and σ_Y of the symmetry transformation of the rectangle in the previous chapter can be expressed as

$$T^{(\mu)}(\sigma_X)T^{(\mu)}(\sigma_Y) = T^{(\mu)}(\sigma_X\sigma_Y), \quad ^\forall \mu \in R(G) \tag{8.19}$$

8.3 Application of Element Rigidity by Symmetry Transformation

When considering the symmetry transformation of various member elements of a finite element, the coordinate transformation matrix based on its symmetry is assembled. Here, we show application examples for basic various element stiffness.

8.3.1 Truss Elements

Consider the symmetry of the axial truss element stiffness

$$K_e = \frac{EA}{\ell} \begin{pmatrix} 1 & -1 \\ -1 & 1 \end{pmatrix}$$

The symmetry between the nodes in this case corresponds to **Fig. 8.1**. That is, the normalized coordinate transformation matrix brought about by the symmetry transformation operation is

$$H_t = \left(h_v^{(+)}, h_v^{(-)} \right) = \begin{pmatrix} -1/\sqrt{2} & 1/\sqrt{2} \\ 1/\sqrt{2} & 1/\sqrt{2} \end{pmatrix} \tag{8.20}$$

Using this to perform the coordinate transformation, we obtain

$$\widetilde{K}_t = H_t^T K_t H_t = \mathrm{diag}\left[\frac{2EA}{\ell},\ 0\right] = \begin{pmatrix} 2 & 0 \\ 0 & 0 \end{pmatrix} \tag{8.21}$$

and the stiffness matrix becomes the stiffness component in the axial compression (tension) direction and the component corresponding to the rigid displacement mode by parallel displacement by symmetry transformation.

8.3.2 Beam Elements

A beam element with 2 degrees of freedom per node, deflection and deflection angle, is a space shared by translational displacement and rotational displacement. The coordinate transformation matrix of this space can be decomposed as

$$h_v^\mu \bigoplus h_\theta^\mu, \quad {}^\forall \mu \in R(\mathrm{D}_2) \tag{8.22}$$

according to Eq.(5.45). This element has the symmetry corresponding to the symmetry transformation of the rotational displacement in **Fig. 8.1** and the translational displacement in **Fig. 8.2**. The coordinate transformation matrices $h_b^{(+)} = \{h_v^{(+)} \bigoplus h_\theta^{(+)}\}$ and $h_b^{(-)} = \{h_v^{(-)} \bigoplus h_\theta^{(-)}\}$ for the irreducible representation μ can be assembled as

$$
\begin{aligned}
H_b &= \left(h_b^{(+)},\ h_b^{(-)}\right) \\
&= \left(h_v^{(+)} \bigoplus h_\theta^{(+)},\ h_v^{(-)} \bigoplus h_\theta^{(-)}\right) \\
&= \begin{pmatrix}
-1/\sqrt{2} & 0 & 1/\sqrt{2} & 0 \\
0 & 1/\sqrt{2} & 0 & -1/\sqrt{2} \\
1/\sqrt{2} & 0 & 1/\sqrt{2} & 0 \\
0 & 1/\sqrt{2} & 0 & 1/\sqrt{2}
\end{pmatrix}
\end{aligned} \tag{8.23}
$$

By using this transformation matrix, the dense 4×4 element stiffness matrix K_b becomes

$$
\begin{aligned}
\widetilde{K}_b &= H_b^T K_b H_b = \frac{2EI}{\ell^3}\left(\begin{array}{cc|cc} 12 & 6\ell & & \\ 6\ell & 3\ell^2 & & \\ \hline & & 0 & 0 \\ & & 0 & \ell^2 \end{array}\right) \\
&= \mathrm{diag.}\left[\widetilde{K}_b^{(+)},\ \widetilde{K}_b^{(-)}\right]
\end{aligned} \tag{8.24}
$$

and is orthogonalized in the irreducible representations $(+), (-)$. Also, in units of irreducible representation, it is commutable as

$$\widetilde{K}_b^\mu = (H_b^\mu)^T K_b H_b^\mu, \quad \mu = (+), (-). \tag{8.25}$$

8.3.3   Diagonalization of the Stiffness Matrix of Rectangular Elements

The symmetry transformation of the plane rectangular element with 2 degrees of freedom per node is shown in **Fig. 8.3** as a combination of two line groups. This group D_2 has four symmetry groups: D_2, C_2, and $D_1 \times 2$ as its subgroups. Of these subgroups, D_2, C_2 correspond to the first irreducible representation, and D_1 corresponds to the second irreducible representation. By assembling the coordinate transformation matrix that corresponds to the symmetry deformation mode shown in **Fig. 8.3** and transforming it as in Eq.(8.25), the dense element stiffness matrix becomes

$$
\widetilde{K}_{\mathrm r} = K_0
\begin{pmatrix}
1 & -\nu & & & & & & \\
-\nu & 1 & & & & & & \\
& & c_1 & -c_1 & & & & \\
& & -c_1 & c_1 & & & & \\
& & & & 0 & 0 & & \\
& & & & 0 & c_2 & & \\
& & & & & & 0 & 0 \\
& & & & & & 0 & c_2
\end{pmatrix}
\tag{8.26}
$$

and becomes a block diagonal form. Here, $K_0 = Et/(1-\nu^2)$, $c_1 = (1-\nu)/2$, $c_2 = (3-\nu)/6$.

If the symmetry of the element is square, the order of the symmetry increases, and the symmetry group is decomposed as

$$
D_2 \longrightarrow \begin{cases} D_4 \\ D_2 \end{cases}, \quad C_2 \longrightarrow \begin{cases} C_4 \\ C_2 \end{cases}
\tag{8.27}
$$

Accordingly, the first irreducible representation μ is also represented as

$$
(+)_{D_2} \longrightarrow \begin{cases} (+,+)_{D_4} \\ (+,-)_{D_4} \end{cases}, \quad (-)_{D_2} \longrightarrow \begin{cases} (-,+)_{D_4} \\ (-,-)_{D_4} \end{cases}
\tag{8.28}
$$

Here, $(\cdot,\cdot)_{D_n}$ represents an irreducible representation with order n. [4] Therefore, the subgroup of the group D_4 has six subgroups: D_4, C_4, D_2, C_2, $D_1 \times 2$. In other words, the element stiffness of the square corresponding to the first irreducible representation can further be decomposed into

$$
K_{\mathrm r}^{(+)_{D_4}} = \mathrm{diag}\left[\widetilde{K}_{\mathrm r}^{(+,+)_{D_4}}, \widetilde{K}_{\mathrm r}^{(+,-)_{D_4}}\right] = K_0\left(1+\nu \,\middle|\, \begin{matrix} \\ 1-\nu \end{matrix}\right),
$$

$$
K_{\mathrm r}^{(-)_{D_4}} = \mathrm{diag}\left[\widetilde{K}_{\mathrm r}^{(-,+)_{D_4}}, \widetilde{K}_{\mathrm r}^{(-,-)_{D_4}}\right] = K_0\left(0 \,\middle|\, \begin{matrix} \\ 1-\nu \end{matrix}\right)
\tag{8.29}
$$

and the number of independent blocks increases. In this way, by increasing the order of symmetry, the number of irreducible representations increases, and the block matrix can be refined.

---

[4] In the other notation, it is assumed as same irreducible representation $\mu \in R(G)$ of group G in the following: $(+,+)_G = (1,1)_G, (+,-)_G = (1,2)_G, (-,+)_G = (1,3)_G, (-,-)_G = (1,4)_G$

8.4 Diagonalization of 3D Solid Elements

8.4.1 Element Stiffness Matrix

We focus on the block diagonalization of the element stiffness matrix for solid elements that have 8 nodes and 3 translational degrees of freedom,[5] giving a total of 24 degrees of freedom, using combinations of line groups for coordinate transformation. As shown in **Fig. 8.4**, if there is 1 degree of freedom per node, there can be 8 combinations of symmetry transformations, and because there are 3 axes directions, there are a total of 24 symmetry transformations. For example, the symmetry transformation $\sigma_Z \sigma_Y \sigma_X$ is constituted by the column vector

$$h(\sigma_Z \sigma_Y \sigma_X) = \frac{1}{\sqrt{8}} \{-1,\ 1,\ 1, -1, -1,\ 1,\ 1, -1\}^{\mathrm{T}} \tag{8.30}$$

As shown in **Fig. 8.6**, there exist 24 types of such transformation vectors. The transformation groups in the figure are classified into six irreducible representations, denoted as $\mu \equiv \{(+), (-), (1), (2), (3), (4)\}$. Each transformation group governs a block after block-diagonalization. Using these transformation groups to transform the 24×24 size element stiffness matrix K_{s}, a block-diagonal form for the solid element emerges, as shown in **Fig. 8.7**.

After the transformation, 14 blocks appear in the stiffness matrix, and each block becomes independent of each other. Each block after block-diagonalization decomposes in accordance with the degree of the irreducible representation as

$$\widetilde{K}_{\mathrm{s}} = \mathrm{diag}\left[\widetilde{K}_{\mathrm{s}}^{(+)}, \widetilde{K}_{\mathrm{s}}^{(-)}, \widetilde{K}_{\mathrm{s}}^{(1)}, \widetilde{K}_{\mathrm{s}}^{(2)}, \widetilde{K}_{\mathrm{s}}^{(3)}, \widetilde{K}_{\mathrm{s}}^{(4)}\right] \tag{8.31}$$

Furthermore, $\widetilde{K}_{\mathrm{s}}^{(1)}$ is as follows:

$$\widetilde{K}_{\mathrm{s}}^{(1)} = \mathrm{diag}\left[\begin{pmatrix} c_2 & c_2 \\ c_2 & c_2 \end{pmatrix}, \begin{pmatrix} c_2 & c_2 \\ c_2 & c_2 \end{pmatrix}, \begin{pmatrix} c_2 & c_2 \\ c_2 & c_2 \end{pmatrix}\right] \tag{8.32}$$

It can be decomposed into three identical blocks. Such blocks are particularly advantageous in numerical computations.

In addition, the left side of **Fig. 8.8** shows the diagonal components of the block diagonal after the element stiffness matrix when the Poisson's ratio is $\nu = 0.3$. The diagonal components within the block show the same value, corresponding to the strain energy amount of the deformation mode for each symmetric transformation. The diagonal components of the block for the same irreducible representation class have the same strain energy, representing the

---

[5]If rotational displacements are considered, this extends to 48 degrees of freedom. In these elements with 48 degrees of freedom, combinations of symmetry can be possible, similar to beam elements. However, here we focus on coordinate transformations by combinations of line groups for translational displacement only.

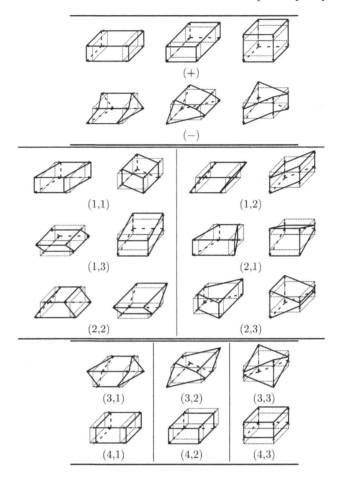

FIGURE 8.6
Transformation groups for cuboid element

energy level of internal energy with respect to strain energy. The right side of **Fig. 8.8** shows the eigenvalues calculation results of each block matrix. After block decomposition, some eigenvalues can be accurately obtained by calculation within the block alone. For example, the block $\widetilde{K}_{s}^{(+)}$ is as follows.

$$\widetilde{K}_{s}^{(+)} = K_0 \begin{pmatrix} 1 & -c_1 & c_1 \\ -c_1 & 1 & -c_1 \\ c_1 & -c_1 & 1 \end{pmatrix} \qquad (8.33)$$

By utilizing the orthogonality of the diagonal blocks, the eigenvalues for this block can easily be obtained as

$$\{(1-c_1)K_0,\ (1-c_1)K_0,\ (1+2c_1)K_0\} \qquad (8.34)$$

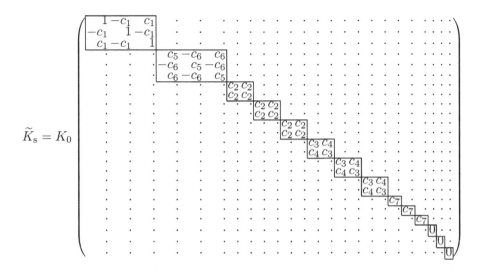

$$K_0 = \frac{(1-\nu)E}{(1+\nu)(1-2\nu)}, \quad c_1 = \frac{\nu}{1-\nu}, \quad c_2 = \frac{1-2\nu}{2(1-\nu)},$$

$$c_3 = (1+c_2)/3, \quad c_4 = c_1/3, \quad c_5 = 2c_2/3, \quad c_6 = c_2/3, \quad c_7 = (1+2c_2)/9.$$

FIGURE 8.7
The matrix form after the block diagonalization of cuboid solid element

This calculation method has the advantage of being able to accurately determine eigenvalues with multiple roots, as in Eq.(8.32).

Regarding the matrix size, there are 45 effective matrix components excluding rigid displacement modes, accounting for 7.8% of the total, and the block $\widetilde{K}_s^{(+)}$ accounts for about 1.6% of the total. The compression ratio of the stiffness matrix for this element and the computational efficiency due to parallel processing are greatly improved. Thus, the utility of symmetry in cubic solid elements is significant, and substantial reductions in the required array capacity and computational time of the conventional stiffness matrix can be expected.

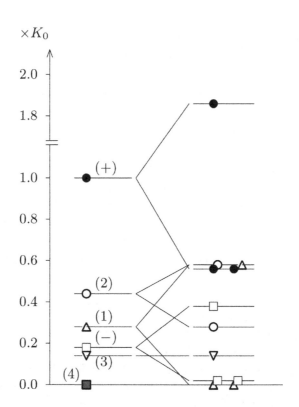

FIGURE 8.8
Strain energy amount and eigenvalues.

A

D_3 Invariant Plate Element

A.1 Transformed Matrix

The nodal displacement vector corresponding to the coordinate transformation matrix consists of 3 nodes and is shown as

$$\boldsymbol{u} = [(\boldsymbol{u}_1)^{\mathrm{T}}, (\boldsymbol{u}_2)^{\mathrm{T}}, (\boldsymbol{u}_3)^{\mathrm{T}}]^{\mathrm{T}}.$$

and $\boldsymbol{u}_i$ is

$$\boldsymbol{u}_i = [u_x, u_y, \theta_z,\ u_z, \theta_x, \theta_y]_i^{\mathrm{T}}.$$

Expressing the coordinate transformation matrix of the equilateral triangular plate element for each irreducible representation,

$$H = [H^{(1,1)}\ \ H^{(1,2)}\ \ H^{(2,1)+}\ \ H^{(2,1)-}]$$

where

$$
H^{(1,1)} =
\begin{pmatrix}
\frac{-1}{\sqrt{3}} & 0 & 0 \\
0 & 0 & 0 \\
0 & 0 & 0 \\
0 & \frac{1}{\sqrt{3}} & 0 \\
0 & 0 & 0 \\
0 & 0 & \frac{1}{\sqrt{3}} \\
\frac{1}{2\sqrt{3}} & 0 & 0 \\
\frac{-1}{2} & 0 & 0 \\
0 & 0 & 0 \\
0 & \frac{1}{\sqrt{3}} & 0 \\
0 & 0 & \frac{-1}{2} \\
0 & 0 & \frac{-1}{2\sqrt{3}} \\
\frac{1}{2\sqrt{3}} & 0 & 0 \\
\frac{1}{2} & 0 & 0 \\
0 & 0 & 0 \\
0 & \frac{1}{\sqrt{3}} & 0 \\
0 & 0 & \frac{1}{2} \\
0 & 0 & \frac{-1}{2\sqrt{3}}
\end{pmatrix},
\quad
H^{(1,2)} =
\begin{pmatrix}
0 & 0 & 0 \\
\frac{1}{\sqrt{3}} & 0 & 0 \\
0 & 0 & \frac{1}{\sqrt{3}} \\
0 & 0 & 0 \\
0 & \frac{-1}{\sqrt{3}} & 0 \\
0 & 0 & 0 \\
\frac{-1}{2} & 0 & 0 \\
\frac{-1}{2\sqrt{3}} & 0 & 0 \\
0 & 0 & \frac{1}{\sqrt{3}} \\
0 & 0 & 0 \\
0 & \frac{1}{2\sqrt{3}} & 0 \\
0 & \frac{-1}{2} & 0 \\
\frac{1}{2} & 0 & 0 \\
\frac{-1}{2\sqrt{3}} & 0 & 0 \\
0 & 0 & \frac{1}{\sqrt{3}} \\
0 & 0 & 0 \\
0 & \frac{1}{2\sqrt{3}} & 0 \\
0 & \frac{1}{2} & 0
\end{pmatrix}
$$

DOI: 10.1201/9781032670386-A

$$H^{(2,1)+} = \left(\begin{array}{cccccc}
\frac{1}{\sqrt{3}} & \frac{-1}{\sqrt{3}} & 0 & 0 & 0 & 0 \\
0 & 0 & 0 & 0 & 0 & 0 \\
0 & 0 & 0 & 0 & 0 & 0 \\
0 & 0 & \frac{2}{\sqrt{6}} & 0 & 0 & 0 \\
0 & 0 & 0 & 0 & 0 & 0 \\
0 & 0 & 0 & \frac{1}{\sqrt{3}} & \frac{-1}{\sqrt{3}} & 0 \\
\hline
\frac{1}{\sqrt{3}} & \frac{1}{2\sqrt{3}} & 0 & 0 & 0 & 0 \\
0 & \frac{1}{2} & 0 & 0 & 0 & 0 \\
0 & 0 & 0 & 0 & 0 & \frac{-1}{\sqrt{2}} \\
0 & 0 & \frac{-1}{\sqrt{6}} & 0 & 0 & 0 \\
0 & 0 & 0 & 0 & \frac{-1}{2} & 0 \\
0 & 0 & 0 & \frac{1}{\sqrt{3}} & \frac{1}{2\sqrt{3}} & 0 \\
\hline
\frac{1}{\sqrt{3}} & \frac{1}{2\sqrt{3}} & 0 & 0 & 0 & 0 \\
0 & \frac{-1}{2} & 0 & 0 & 0 & 0 \\
0 & 0 & 0 & 0 & 0 & \frac{1}{\sqrt{2}} \\
0 & 0 & \frac{-1}{\sqrt{6}} & 0 & 0 & 0 \\
0 & 0 & 0 & 0 & \frac{1}{2} & 0 \\
0 & 0 & 0 & \frac{1}{\sqrt{3}} & \frac{1}{2\sqrt{3}} & 0
\end{array}\right)$$

$$H^{(2,1)-} = \left(\begin{array}{cccccc}
0 & 0 & 0 & 0 & 0 & 0 \\
\frac{1}{\sqrt{3}} & \frac{-1}{\sqrt{3}} & 0 & 0 & 0 & 0 \\
0 & 0 & 0 & 0 & 0 & \frac{2}{\sqrt{6}} \\
0 & 0 & 0 & 0 & 0 & 0 \\
0 & 0 & 0 & \frac{1}{\sqrt{3}} & \frac{-1}{\sqrt{3}} & 0 \\
0 & 0 & 0 & 0 & 0 & 0 \\
\hline
0 & \frac{-1}{2} & 0 & 0 & 0 & 0 \\
\frac{1}{\sqrt{3}} & \frac{1}{2\sqrt{3}} & 0 & 0 & 0 & 0 \\
0 & 0 & 0 & 0 & 0 & \frac{-1}{\sqrt{6}} \\
0 & 0 & \frac{-1}{\sqrt{2}} & 0 & 0 & 0 \\
0 & 0 & 0 & \frac{1}{\sqrt{3}} & \frac{1}{2\sqrt{3}} & 0 \\
0 & 0 & 0 & 0 & \frac{1}{2} & 0 \\
\hline
0 & \frac{1}{2} & 0 & 0 & 0 & 0 \\
\frac{1}{\sqrt{3}} & \frac{1}{2\sqrt{3}} & 0 & 0 & 0 & 0 \\
0 & 0 & 0 & 0 & 0 & \frac{-1}{\sqrt{6}} \\
0 & 0 & \frac{1}{\sqrt{2}} & 0 & 0 & 0 \\
0 & 0 & 0 & \frac{1}{\sqrt{3}} & \frac{1}{2\sqrt{3}} & 0 \\
0 & 0 & 0 & 0 & \frac{-1}{2} & 0
\end{array}\right)$$

A.2 Element Stiffness Matrix

The element stiffness matrix for triangular plate bending is expressed as

$$K_e = \begin{pmatrix} K_e^{11} & K_e^{12} & K_e^{13} \\ & K_e^{22} & K_e^{23} \\ \text{Symm.} & & K_e^{33} \end{pmatrix}.$$

where

$$K_e^{11} = \begin{pmatrix} \frac{64D_P}{3} & 0 & 0 & 0 & 0 & 0 \\ 0 & \frac{112D_P}{15} & 0 & 0 & 0 & 0 \\ 0 & 0 & D_T & 0 & 0 & 0 \\ 0 & 0 & 0 & \frac{128D_B}{9L^2} & 0 & \frac{-44D_B}{5\sqrt{3}L} \\ 0 & 0 & 0 & 0 & \frac{208D_B}{135} & 0 \\ 0 & 0 & 0 & \frac{-44D_B}{5\sqrt{3}L} & 0 & \frac{46D_B}{15} \end{pmatrix}$$

$$K_e^{12} = \begin{pmatrix} \frac{-32D_P}{3} & \frac{48D_P}{5\sqrt{3}} & 0 & 0 & 0 & 0 \\ \frac{56D_P}{5\sqrt{3}} & \frac{-56D_P}{15} & 0 & 0 & 0 & 0 \\ 0 & 0 & \frac{-D_T}{2} & 0 & 0 & 0 \\ 0 & 0 & 0 & \frac{-64D_B}{9L^2} & \frac{-32D_B}{9L} & \frac{28\sqrt{3}D_B}{45L} \\ 0 & 0 & 0 & \frac{-122D_B}{45L} & \frac{-139D_B}{270} & \frac{-59\sqrt{3}D_B}{270} \\ 0 & 0 & 0 & \frac{22D_B}{5\sqrt{3}L} & \frac{9D_B}{10\sqrt{3}} & \frac{-13D_B}{30} \end{pmatrix}$$

$$K_e^{13} = \begin{pmatrix} \frac{-32D_P}{3} & \frac{-48D_P}{5\sqrt{3}} & 0 & 0 & 0 & 0 \\ \frac{-56D_P}{5\sqrt{3}} & \frac{-56D_P}{15} & 0 & 0 & 0 & 0 \\ 0 & 0 & \frac{-D_T}{2} & 0 & 0 & 0 \\ 0 & 0 & 0 & \frac{-64D_B}{9L^2} & \frac{32D_B}{9L} & \frac{28\sqrt{3}D_B}{45L} \\ 0 & 0 & 0 & \frac{122D_B}{45L} & \frac{-139D_B}{270} & \frac{59\sqrt{3}D_B}{270} \\ 0 & 0 & 0 & \frac{22D_B}{5\sqrt{3}L} & \frac{-9D_B}{10\sqrt{3}} & \frac{-13D_B}{30} \end{pmatrix}$$

$$K_e^{22} = \begin{pmatrix} \frac{164D_P}{15} & \frac{-52D_P}{5\sqrt{3}} & 0 & 0 & 0 & 0 \\ \frac{-52D_P}{5\sqrt{3}} & \frac{268D_P}{15} & 0 & 0 & 0 & 0 \\ 0 & 0 & D_T & 0 & 0 & 0 \\ 0 & 0 & 0 & \frac{128D_B}{9L^2} & \frac{22D_B}{5L} & \frac{22\sqrt{3}D_B}{15L} \\ 0 & 0 & 0 & \frac{22D_B}{5L} & \frac{145D_B}{54} & \frac{103\sqrt{3}D_B}{270} \\ 0 & 0 & 0 & \frac{22\sqrt{3}D_B}{15L} & \frac{103\sqrt{3}D_B}{270} & \frac{173D_B}{90} \end{pmatrix}$$

$$K_e^{23} = \begin{pmatrix} \frac{-4D_P}{15} & \frac{-4D_P}{5\sqrt{3}} & 0 & 0 & 0 & 0 \\ \frac{4D_P}{5\sqrt{3}} & \frac{-212D_P}{15} & 0 & 0 & 0 & 0 \\ 0 & 0 & \frac{-D_T}{2} & 0 & 0 & 0 \\ 0 & 0 & 0 & \frac{-64D_B}{9L^2} & \frac{38D_B}{45L} & \frac{-94\sqrt{3}D_B}{45L} \\ 0 & 0 & 0 & \frac{-38D_B}{45L} & \frac{-53D_B}{135} & \frac{-7\sqrt{3}D_B}{27} \\ 0 & 0 & 0 & \frac{-94\sqrt{3}D_B}{45L} & \frac{7\sqrt{3}D_B}{27} & \frac{-5D_B}{9} \end{pmatrix}$$

$$K_e^{33} = \begin{pmatrix} \frac{164D_P}{15} & \frac{52D_P}{5\sqrt{3}} & 0 & 0 & 0 & 0 \\ \frac{52D_P}{5\sqrt{3}} & \frac{268D_P}{15} & 0 & 0 & 0 & 0 \\ 0 & 0 & D_T & 0 & 0 & 0 \\ 0 & 0 & 0 & \frac{128D_B}{9L^2} & \frac{-22D_B}{5L} & \frac{22\sqrt{3}D_B}{15L} \\ 0 & 0 & 0 & \frac{-22D_B}{5L} & \frac{145D_B}{54} & \frac{-103\sqrt{3}D_B}{270} \\ 0 & 0 & 0 & \frac{22\sqrt{3}D_B}{15L} & \frac{-103\sqrt{3}D_B}{270} & \frac{173D_B}{90} \end{pmatrix}$$

$$D_P = \frac{E\,t}{1-\nu^2}\frac{A}{L^2} \quad, \quad D_T = \alpha\,E\,t\,A \quad, \quad D_B = \frac{E\,t^3}{12(1-\nu^2)}\frac{A}{L^2} \quad, \text{and} \quad \nu = 0.3\ .$$

A.3   Element Block-diagonalization Matrix

The block-diagonalized element stiffness matrix $\widetilde{K}_e$ is shown

$$\widetilde{K}_e = \begin{pmatrix} \widetilde{K}_e^{(1,1)} & O & O & O \\ & \widetilde{K}_e^{(1,2)} & O & O \\ \text{Symm.} & & \widetilde{K}_e^{(2,1)+} & O \\ & & & \widetilde{K}_e^{(2,1)-} \end{pmatrix}.$$

where

$$\widetilde{K}_e^{(1,1)} = \begin{pmatrix} \frac{208D_P}{5} & 0 & 0 \\ 0 & 0 & 0 \\ 0 & 0 & \frac{13D_B}{5} \end{pmatrix}$$

$$\widetilde{K}_e^{(1,2)} = \begin{pmatrix} 0 & 0 & 0 \\ 0 & \frac{7D_B}{5} & 0 \\ 0 & 0 & 0 \end{pmatrix}$$

$$\widetilde{K}_e^{(2,1)+} = \begin{pmatrix} 0 & 0 & 0 & 0 & 0 & 0 \\ 0 & \frac{112D_P}{5} & 0 & 0 & 0 & 0 \\ 0 & 0 & \frac{64D_B}{3L^2} & -\frac{76D_B}{15\sqrt{6}L} & \frac{64D_B}{3\sqrt{6}L} & 0 \\ 0 & 0 & -\frac{76D_B}{15\sqrt{6}L} & \frac{61D_B}{45} & -\frac{38D_B}{45} & 0 \\ 0 & 0 & \frac{64D_B}{3\sqrt{6}L} & -\frac{38D_B}{45} & \frac{32D_B}{9} & 0 \\ 0 & 0 & 0 & 0 & 0 & \frac{3D_T}{2} \end{pmatrix}$$

$$\widetilde{K}_e^{(2,1)-} = \begin{pmatrix} 0 & 0 & 0 & 0 & 0 & 0 \\ 0 & \frac{112D_P}{5} & 0 & 0 & 0 & 0 \\ 0 & 0 & \frac{64D_B}{3L^2} & \frac{76D_B}{15\sqrt{6}L} & \frac{64D_B}{3\sqrt{6}L} & 0 \\ 0 & 0 & \frac{76D_B}{15\sqrt{6}L} & \frac{61D_B}{45} & \frac{38D_B}{45} & 0 \\ 0 & 0 & \frac{64D_B}{3\sqrt{6}L} & \frac{38D_B}{45} & \frac{32D_B}{9} & 0 \\ 0 & 0 & 0 & 0 & 0 & \frac{3D_T}{2} \end{pmatrix}$$

B

D_2 Invariant Bending Plate Element

B.1 D_2 Bending Stiffness matrix in an Element

The rectangular plate bending element stiffness matrix of the ACM (Adini, Clough and Melosh) [45] incompatible element for isotropic homogeneous material is expressed as follows.

$$
K_e = \frac{Et^3}{90ab(1-\nu^2)}
\begin{pmatrix}
k_e^{11} & k_e^{12} & k_e^{13} & k_e^{14} \\
 & k_e^{22} & k_e^{23} & k_e^{24} \\
 & & k_e^{33} & k_e^{34} \\
\text{Symm.} & & & k_e^{44}
\end{pmatrix}
$$

where

$$
k_e^{11} =
\begin{pmatrix}
3B_8 & -3B_1 & -3B_1 \\
 & 8B_3 & 30\nu \\
\text{Symm.} & & 8B_3
\end{pmatrix}, \quad
k_e^{12} =
\begin{pmatrix}
6B_3 & -3B_4 & -12B_5 \\
3B_4 & 2B_6 & 0 \\
-12B_5 & 0 & 4B_7
\end{pmatrix}
$$

$$
k_e^{13} =
\begin{pmatrix}
-3B_7 & -3B_2 & -3B_2 \\
3B_2 & 2B_3 & 0 \\
3B_2 & 0 & 2B_3
\end{pmatrix}, \quad
k_e^{14} =
\begin{pmatrix}
-6B_3 & -12B_5 & -3B_4 \\
-12B_5 & 4B_7 & 0 \\
3B_4 & 0 & 2B_6
\end{pmatrix}
$$

$$
k_e^{22} =
\begin{pmatrix}
3B_8 & 3B_1 & -3B_1 \\
 & 8B_3 & -30\nu \\
\text{Symm.} & & 8B_3
\end{pmatrix}, \quad
k_e^{23} =
\begin{pmatrix}
-6B_3 & 12B_5 & -3B_4 \\
12B_5 & 4B_7 & 0 \\
3B_4 & 0 & 2B_6
\end{pmatrix}
$$

$$
k_e^{24} =
\begin{pmatrix}
-3B_7 & 3B_2 & -3B_2 \\
-3B_2 & 2B_3 & 0 \\
3B_2 & 0 & 2B_3
\end{pmatrix}, \quad
k_e^{33} =
\begin{pmatrix}
3B_8 & 3B_1 & 3B_1 \\
 & 8B_3 & 30\nu \\
\text{Symm.} & & 8B_3
\end{pmatrix}
$$

$$
k_e^{34} =
\begin{pmatrix}
-6B_3 & 3B_4 & -12B_5 \\
-3B_4 & 2B_6 & 0 \\
12B_5 & 0 & 4B_7
\end{pmatrix}, \quad
k_e^{44} =
\begin{pmatrix}
3B_8 & -3B_1 & 3B_1 \\
 & 8B_3 & -30\nu \\
\text{Symm.} & & 8B_3
\end{pmatrix}
$$

where $B_1 = 11+4\nu$, $B_2 = 4+\nu$, $B_3 = 6-\nu$, $B_4 = 11-\nu$, $B_5 = 1-\nu$, $B_6 = 6+\nu$, $B_7 = 3+2\nu$, $B_8 = 27-2\nu$.

DOI: 10.1201/9781032670386-B

B.2 Block Diagonalized D_2 Invariant Bending Stiffness Matrix in a Rectangle Plate Element

The block-diagonalized rectangular plate bending element stiffness matrix is expressed as:

$$\widetilde{K}_e = \frac{Et^3}{90ab(1-\nu^2)}\mathrm{diag}[\widetilde{K}_e^{(1,1)D_2}, \widetilde{K}_e^{(1,2)D_2}, \widetilde{K}_e^{(2,1)^+_{D_2}}, \widetilde{K}_e^{(2,1)^-_{D_2}}]$$

where the components for each irreducible representation are following;

$$\widetilde{K}_e^{(1,1)D_2} = \begin{pmatrix} 3(5B_9 - 14\nu + 4) & 15B_{10} & -15B_{10}\sqrt{2} \\ & 3(5B_9 + 6\nu + 4) & -3(5B_9 + 6\nu + 4) \\ & \mathrm{Symm.} & 6(5B_9 - 4\nu + 14) \end{pmatrix}$$

$$\widetilde{K}_e^{(1,2)D_2} = \begin{pmatrix} 15(B_{10} - 2\nu) & 15B_{10} \\ \mathrm{Symm.} & 15(B_{10} + 2\nu) \end{pmatrix}$$

$$\widetilde{K}_e^{(2,1)^+_{D_2}} = \begin{pmatrix} 90\beta^{-2} & -30\nu & -90\beta^{-2} \\ & 10(2 - 2\nu + \beta^2) & 30\nu \\ \mathrm{Symm.} & & 90\beta^{-2} \end{pmatrix}$$

$$\widetilde{K}_e^{(2,1)^-_{D_2}} = \begin{pmatrix} 90\beta^2 & -30\nu & -90\beta^2 \\ & 10(2 - 2\nu + \beta^{-2}) & 30\nu \\ \mathrm{Symm.} & & 90\beta^2 \end{pmatrix}$$

Also, the aspect ratio of the strip width is $\beta = b/a$, and the variables $B_9 = \beta^{-2} + \beta^2$, $B_{10} = \beta^{-2} - \beta^2$.

B.3 Block Diagonalized D_4 Invariant Bending Stiffness Matrix in a Square Plate Element

The block-diagonalized square plate bending element stiffness matrix is following;

$$\widetilde{K}_e = \frac{Et^3}{90ab(1-\nu^2)}\mathrm{diag}[\widetilde{K}_e^{(1,1)D_4}, \widetilde{K}_e^{(1,2)D_4}, \widetilde{K}_e^{(1,3)D_4}, \widetilde{K}_e^{(1,4)D_4}, \widetilde{K}_e^{(2,1)^+_{D_4}}, \widetilde{K}_e^{(2,1)^-_{D_4}}]$$

where

$$
\widetilde{K}_e^{(1,1)_{D_4}} = 42B_5
$$

$$
\widetilde{K}_e^{(1,2)_{D_4}} = \begin{pmatrix} 6(7+3\nu) & -6(7+3\nu)\sqrt{2} \\ & 24B_3 \end{pmatrix}
$$

$$
\widetilde{K}_e^{(1,3)_{D_4}} = 30B_5
$$

$$
\widetilde{K}_e^{(1,4)_{D_4}} = 30(1+\nu)
$$

$$
\widetilde{K}_e^{(2,1)_{D_4}^+} = \begin{pmatrix} 6(7+3\nu) & -6(7+3\nu)\sqrt{2} \\ & 24B_3 \end{pmatrix}
$$

$$
\widetilde{K}_e^{(2,1)_{D_4}^+} = \begin{pmatrix} 90 & -30\nu & -90 \\ & 10(3-2\nu) & 30\nu \\ \text{Symm.} & & 90 \end{pmatrix}
$$

$$
\widetilde{K}_e^{(2,1)_{D_4}^-} = \widetilde{K}_e^{(2,1)_{D_4}^+}
$$

Bibliography

[1] Zak, A.R.,Craddock, J.N. and Drysdale, W.H.: Approximate Finite Element Method of Stress Analysis of Non-Axisymmetric Configurations, *Computer & Struct.*, 9, 201-206, 1979. DOI:10.1016/0045-7949(78)90139-6

[2] Nagamatsu, A. and Okuma, M.: Partial Structure Synthesis Method, Baifukan, 1991.

[3] Egeland, O. and Araldsen, H.: A general purpose finite element method program, *Computers & Structures*, 4, 41-68, 1974. DOI:10.1016/0045-7949(74)90076-5

[4] Torabi, J. and Ansari, R.: A higher-order isoparametric superelement for free vibration analysis of functionally graded shells of revolution, *Thin-Walled Structures*, 133, 169–179, December 2018. DOI:10.1016/j.tws.2018.09.040

[5] Tallarico, D., Hannema, G., Miniaci, M., Bergamini, A., Zempa, A., Van Damme, B.: Superelement modelling of elastic metamaterials: Complex dispersive properties of three-dimensional structured beams and plates, *Journal of Sound and Vibration*, **484**(13), October 2020, 115499. DOI:10.1016/j.jsv.2020.115499

[6] Superelement User's Guide, Siemens https://docs.plm.automation.siemens.com/data_services/resources/nxnastran/10/help/en_US/tdocExt/pdf/super.pdf, Accessed 12.16.2023

[7] Clough, R.W. and Rashid, Y.: Finite Element Analysis of Axi-symmetric Solids, *Proc.ASCE*,91,EM1, 71-85, 1965. DOI:10.1061/JMCEA3.0000585

[8] Zienkiwicz, O.Z., Irons, B.M., Ergatoudis, J., Ahmed, S. and Scott, F.C.: Isoparametric and Associated Element Families for Two-and Three-dimensional Analysis Finite Element Methods, edited by I.Holland and K.Bell, Tapir, 1969.

[9] Crose, J.G.: Stress Analysis of Axisymmetric Solids with Axisymmetric Properties, *AIAA Journal*, 10, 866-871, 1972. DOI:10.2514/3.50238

179

[10] Sattinger, D.H.: Group Theoretic Methods in Bifurcation Theory, Lecture Notes in Mathematics (LNM, volume 762), Springer, 1979.

[11] Thompson, J.M.T and Hunt, G.W.: A General Theory of Elastic Stability, John Wiley, 1973.

[12] Baumslag, B. and Chandler, B.: Theory and Problems of G-Group Theory, Outline Series in Mathematics, McGraw-Hill, New York, 1968.

[13] Cotton, F.A.: Chemical Applications of Group Theory, 2nd edition, John Wiley, 1971.

[14] Kettle, S.F.A.: Symmetry and Structure, John Wiley, Chichester, 1985.

[15] Van der Waerden, B.L.: Group Theory and Quatum Mechanics, Grundlehren der Mathematischen Wissenschaften, 214, Springer, 1980.

[16] Fujii, H. and Yamaguti, M.: Structure of singularities and its numerical realization in nonlinear elasticity, *Journal of Mathematics of Kyoto University*, 20, 498-590, 1980.

[17] Zloković, G.: Group Theory and G-vector Spaces in Vibrations, Stability and Statics of Structures, *ICS, Beograd*, (In English and Serbo-Croatian), 1973.

[18] Golubitsky, M. and Schaeffer, D.G.: Singularities and Groups in Bifurcation Theory, Vol.1, Springer, 1985.

[19] Golubitsky, M., Stewart, I. and Schaeffer, D.G.: Singularities and Groups in Bifurcation Theory, Vol.2, Springer, 1988.

[20] Bossavit, A.: Symmetry, groups, and boundary value problems. A progressive introduction to noncommutative harmonic analysis of partial differential equations in domains with geometrical symmetry, *Computer Methods in Applied Mechanics and Engineering*, **56**(2), 167-215, 1986. DOI:10.1016/0045-7825(86)90119-2

[21] Zloković, G.: Group Theory and G-vector Spaces in Structural Analysis, John Wiley and Sons, 1989.

[22] Healey, T.J.: A group theoretic approach to computational bifurcation problems with symmetry, *Computer Methods in Applied Mechanics and Engineering*, 67, 257-295, 1988. DOI:10.1016/0045-7825(88)90049-7

[23] Healey, T.J. and Treacy, J.A.: Exact block diagonalization of large eigenvalue problems for structures with symmetry, *International Journal for Numerical Methods in Engineering*, 31, 265-285, 1991. DOI:10.1002/nme.1620310205

[24] Dinkevich, S.: The spectral method of calculation of symmetric structures of finite size, *Transactions of the Canadian Society for Mechanical Engineering*, 8(4), 185-194, 1984. DOI:10.1139/tcsme-1984-0028

[25] Dinkevich, S.: The fast method of block elimination for the solution of large regular mechanical structures, *Transactions of the Canadian Society for Mechanical Engineering*, 10(2), 91-98, 1986. DOI:10.1139/tcsme-1986-0011

[26] Dinkevich, S.: Finite symmetric systems and their analysis, *International Journal of Solids and Structures*, 27(10), 1215-1253, 1991. DOI:10.1016/0020-7683(91)90160-H

[27] Murota, K. and Ikeda, K.: Computational use of group theory in bifurcation analysis of symmetric structures, *SIAM Journal on Statistical and Scientific Computing*, 12(2), 273-297, 1991. DOI:10.1137/0912016

[28] Murota, K. and Ikeda, K.: On random imperfections for structures of regular-polygonal symmetry, *SIAM Journal on Applied Mathematics*, 52(6), 1780-1803, 1992. DOI:10.1137/0152102

[29] Ikeda, K., Murota, K. and Fujii, H.: Bifurcation hierarchy of symmetric structures, *International Journal of Solids and Structures*, 27(12), 1551-1573, 1991. DOI:10.1016/0020-7683(91)90077-S

[30] Ikeda, K. and Murota, K.: Bifurcation analysis of symmetric structures using block-diagonalization, *Computer Methods in Applied Mechanics and Engineering*, 86(2), 215-243, 1991. DOI:10.1016/0045-7825(91)90128-S

[31] Ikeda, K., Ario, I. and Torii, K.: Block-diagonali-zation analysis of symmetric plates, *International Journal of Solids and Structures*, 29(22), 2779-2793, 1992. DOI:10.1016/0020-7683(92)90118-D

[32] Zingoni, A.: An efficient computational scheme for the vibration analysis of high-tension cable nets, *J. of Sound and Vibration*, 189, 55-79, 1996. DOI:10.1006/jsvi.1996.0005

[33] Zingoni, A.: Group-theoretical applications in solid and structural mechanics: a review, chapter 12. In: Topping, B.H.V., Bittnar, Z., (eds.) Computational Structures Technology, Saxe-Coburg Publications, Stirlingshire, 2002.

[34] Zingoni, A.: On the symmetries and vibration modes of layered space grids, *Eng. Struct.* 27, 629-638, 2005. DOI:10.1016/j.engstruct.2004.12.004

[35] Zingoni, A.: On group-theoretic computation of natural frequencies for spring-mass dynamic systems with rectilinear motion, *Commun. Numer. Methods Eng.* 24, 973-987, 2008. DOI:10.1002/cnm.1003

[36] Zhang, J.Y., Guest, S.D., Ohsaki, M.: Symmetric prismatic tensegrity structures. Part I: Configuration and stability, *Int. J. Solids Struct.* 46, 1-14, 2009. DOI:10.1016/j.ijsolstr.2008.08.032

[37] Kaveh, A., Nikbakht,M.: Block diagonalization of Laplacian matrices of symmetric graphs via group theory, *Int. J. Numer. Methods Eng.* 69, 908-947, 2007. DOI:10.1002/nme.1794

[38] Kaveh, A., Nikbakht, M.: Improved group-theoretical method for eigenvalue problems of special symmetric structures, using graph theory, *Adv. Eng. Softw.*, 2009. DOI:10.1016/j.advengsoft.2008.12.003

[39] Kaveh, A., Nikbakht, M.: Stability analysis of hyper symmetric skeletal structures using group theory, *ActaMech.* 200, 177-197, 2008. DOI:10.1007/s00707-008-0022-x

[40] Zingoni, A.: Vibration Analysis and Structural Dynamics for Civil Engineers, Essentials and Group-Theoretic Formulations, CRC Press, Taylor & Francis Group, 2015.

[41] Hunt, G.W. and Ario, I.: Twist buckling and the foldable cylinder: an exercise in origami, *International Journal of Non-Linear Mechanics*, 40(6), 833-843, 2005. DOI:10.1016/j.ijnonlinmec.2004.08.011

[42] Kresling, B.: Folded Tubes as Compared to Kikko ("Tortoise-Shell") Bamboo, *Origami3, Third International Meeting of Origami Science, Mathematics and Education (2001 Asilomar, California)(Thomas Hull, editor) AK Peters, Natick, Massachusetts*, 197-207, 2002.

[43] Ario, I. and Nakazawa, M.: Optimization problem of cylindrical structure applied the geometrical structure of a bamboo, *Symposium on System Optimization in Japan Society of Civil Engineers (JSCE)*, Vol.43A, 1997.

[44] Ario, I., Morita, C., Suyma, H., Sato, E. and Fuji, K.: Mechanical Considerations of the Laminated Composite Structure Modeled on the Anisotropic Organization of a Bamboo, (Bioengineering of Biological and Medical Materials), *Transactions of the Japan Society of Mechanical Engineers Series A*, **69**(677), 148-153, 2003. DOI:10.1299/kikaia.69.148

[45] Clough, R.W. and Tocher, J.L.: Finite element stiffness matrices for analysis of plate bending, Conference on Matrix Methods in Structural Mechanics, Wright-Patterson Air Force Base, Ohio, 1965.

[46] Ario, I., Ikeda, K. and Murota, K.: Block-diagonalization method for symmetric structures with rotational displacements, *Journal of Structural Mechanics and Earthquake Engineering, JSCE*, No.483/I-27, 27-36, 1994. DOI:10.2208/jscej.1994.489_27

[47] Ario, I., Ikeda, K. and Torii, K.: Block-diagonalization method for dynamic problems of damped symmetric structures, *Journal of Structural Mechanics and Earthquake Engineering, JSCE*, 41A, 1995.

[48] Chikahiro, Y., Ario, I., Pawłowski, P., Graczykowski, C., Nakazawa, M., Holnicki-Szulc, J., Ono, S.: Dynamics of the scissors-type Mobile Bridge, *Procedia Engineering*, 199, 2919-2924, 2017. DOI:10.1016/j.proeng.2017.09.339

[49] Zawidzki, M. and Nagakura, T.: Foldable Truss-Z module, 16th International Conference on Geometry And Graphics, 4–8 August, Innsbruck, Austria, 2014.

[50] Zawidzki, M.: Discrete Optimization in Architecture: Building Envelope, SpringerBriefs in Architectural Design and Technology, 1st ed. 2017 Edition, 2017.

[51] Zawidzki, M.: Discrete Optimization in Architecture: Extremely Modular Systems, SpringerBriefs in Architectural Design and Technology, 1st ed. 2017 Edition, 2017.

[52] Zawidzki, M. : Discrete Optimization in Architecture: Architectural & Urban Layout, SpringerBriefs in Architectural Design and Technology, 1st ed. 2016 Edition, 2016.

[53] Oguni, R., Murata, T., Miyoshi, T., Dongara, J.J., Hasegawa, H.: Matrix Calculation Software WS, Supercomputer, Parallel Computer, Maruzen, 1991.

[54] Ezawa, H., and Shima, K.: Groups and Representations, Iwanami Course Applied Mathematics 8, Iwanami Shoten, 1994.

[55] Serre, J.P. : Linear Representations of Finite Groups, Springer, 1977.

[56] Marks, R.W.: The Dymaxion World of Buckminster Fuller, Reinhold, New York, 1960.

[57] Fuller, R.B.: Synergetics, Macmillan, New York, 1975.

[58] Edmondson, A.C.: A Fuller Explanation, Van Nostrand Reinhold, New York, 1992.

[59] Kroto, H.W., Heath, J.R., O'Brien, S.C., Curl, R.F. and Smally, R.E.: C_{60} : Buckminsterfullerene, *Nature*, **318**, 162-3, 1985.

[60] Chancy, C.C. and O'Brien M.C.M.: The Jahn-Teller Effect in C_{60} and Other Icosahedral Complexes, Princeton University Press, 1997.

[61] Ario, I., Ikeda, K. and Murota, K.: Block-diagonalization method for symmetric structures with rotational displacements, *Journal of Structural Mechanics and Earthquake Engineering, JSCE*, No.489/I-27, 27-36, 1994.

[62] Ario, I., Fujii, K. and Satoh, M.: Numerical efficiency of paralell cholesky decomposition for the stiffness matrix of symmetric structures by block-diagonalization method, *Structural engineering in JSCE*, Vol.44A, 269-274,(in Japanese) 1998.

[63] Ario, I., Fujii, K. and Sato, M.: Fast Iterative SOR Method by Parallel Computing for the Quasi-Diagonal Block Stiffness Matrix, *Structural engineering in JSCE*, Vol.44A, (in Japanese) 1998.

[64] Yunus, A.M., Saigal, S. and Cook, R.D.: On improved hybrid finite elements with rotational degrees of freedom, *International Journal for Numerical Methods in Engineering*, 28, 785-800, 1989. DOI:10.1002/nme.1620280405

[65] Ibrahimbegovic, A. and Wilson, E.L.: Thick shell and solid finite elements with independent rotation fields, *International Journal for Numerical Methods in Engineering*, 31, 1393-1414, 1991. DOI:10.1002/nme.1620310711

[66] Tian, Z.S., Liu, J.S. and Fang, B.: Stress analyses of solids with rectangular holes by 3-D special hybrid stress elements, *Structural Engineering and Mechanics*, 3(2), 193-199, 1995. DOI:10.12989/sem.1995.3.2.193

[67] Sze, K.Y., Soh, A.K. and Sim, Y.S.: Solid elements with rotational dofs by explicit hybrid stabilization, *International Journal for Numerical Methods in Engineering*, 39, 2987-3005, 1996. DOI:10.1002/(SICI)1097-0207(19960915)39:17<2987::AID-NME986>3.0.CO;2-H

[68] Ario, I.: The Block-Diagonalization Method Applied to the Stiffness Matrix of Solid Finite Elements by Group Product, *Journal of Applied Mechanics in JSCE*, Vol.1 (in Japanese) 1998. https://doi.org/10.2208/journalam.1.223

[69] Ikeda, K., Murota, K. and Nakano, M.: Echelon modes in uniform materials, *International Journal of Solids and Structures*, 31(19), 2709-2733, 1994. DOI:10.1016/0020-7683(94)90226-7

[70] Fujii, H. and Yamaguti, M.: Structure of singularities and its numerical realization in nonlinear elasticity, *J. Math. Kyoto Univ.*, 20, 498-590, 1980. DOI:10.1215/kjm/1250522212

[71] Ikeda,. K., Nakazawa, M. and Wachi, S. : Degeneration of bifurcation hierarchy of a rectangular plate due to boundary conditions, *J. Struct. Mech. Earthquake Eng., JSCE*, No.507/I-30, 65-75, 1995. https://doi.org/10.2208/jscej.1995.507_65

[72] Ario, I., Yamashita, T., Chikahiro, Y., Nakazawa, M., Krzysztof, F., Graczykowski, C., Pawłowski, P.: Structural Analysis of a Scissor Structure, *Bulletin of the Polish Academy of Sciences, Technical Sciences*, 2020.12. DOI:10.24425/bpasts.2020.134623

[73] Lewiński,T., Sokół,T., Graczykowski, C.: Michell Structures, Springer, 2018.

[74] Ario, I., Nakazawa, M., Tanaka, Y., Tanikura, I., Ono S.: Development of a prototype deployable bridge based on origami skill, *Automation in Construction*, 32, 104-111, 2013. DOI:10.1016/j.autcon.2013.01.012

[75] Turner, M.J., Clough, R.W., Martin, H.C., Topp, L.J. : Stiffness and Deflection Analysis of Complex Structures, *Int. Journal of the Aeronautical Sciences*, 23(9), 805-823, 1956. DOI:10.2514/8.3664

[76] Ikeda, K. and Murota, K.: Imperfect Bifurcation in Structures and Materials, Engineering Use of Group-Theoretic Bifurcation Theory, Springer-Verlag, New York, 2002.

[77] Kovács, F.: Extended truss theory with simplex constraints, *International Journal of Solids and Structures*, 48(3-4), 472-482, 2011. DOI:10.1016/j.ijsolstr.2010.10.014

[78] Sun, C.T. and Vaidya, R.S.: Prediction of composite properties from a representative volume element, *Int. Composites Science and Technology*, 56(2), 171-179, 1997. DOI:10.1016/0266-3538(95)00141-7

[79] Ario, I. and Nakazawa, M.: Analysis of multiple bifurcation behavior for periodic structures, *Proc. of 4th Polish Congress of Mechanics and 23rd Int. Conference on Computer Methods in Mechanic*, Krakow, Poland, 2019.

[80] Ario, I. and Nakazawa,M.: Analysis of Multiple Bifurcation Behaviour for Periodic Structures, *Archives of Mechanics*, **72**(4), 1-24, 2020. DOI:10.24423/aom.3433

Index

Printed in the United States
by Baker & Taylor Publisher Services